Point-of-Care Testing for Managers and Policymakers

Point-of-Care Testing for Managers and Policymakers:
From Rapid Testing to Better Outcomes

Christopher P. Price, PhD, FRCPath, FACB
Visiting Professor in Clinical Biochemistry
University of Oxford
Oxford
United Kingdom

Andrew St John, PhD, MAACB
Consultant
ARC Consulting
Perth
Western Australia

1850 K Street, NW, Suite 625
Washington, DC 20006

1 2 3 4 5 6 7 8 9 0 EB 08 07 06

Printed in the United States of America

Library of Congress Cataloging-in-Publication Data

Price, Christopher P.
   Point-of-care testing for managers and policymakers : from rapid testing to better outcome / Christopher P. Price, Andrew St. John.
      p. ; cm.
   Includes bibliographical references and index.
   ISBN 1-59425-051-0
   1. Point-of-care testing—Evaluation.   2. Outcome assessment (Medical care)   I. St. John, Andrew   II. Title.
   [DNLM: 1. Point-of-Care Systems.   2. Clinical Laboratory Techniques.   3. Laboratory Techniques and Procedures.   4. Outcome Assessment (Health Care)   WX 162 P945p 2006]

   RC71.7.P75 2006
   616.07'5—dc22

                                                                                      2006013164

Contents

Preface

Point-of-care testing (POCT) continues to be at the forefront of discussions among laboratory medicine professionals, clinicians, and other caregivers, and also among managers and policymakers within the field of healthcare. This is because technological innovation is routinely delivering new and improved devices for performing diagnostic tests with apparently increasing simplicity as well as better analytical performance and shorter analysis times. More recently, the emphasis in this discussion has shifted to the desire for a more patient-centered approach to care, including treating patients closer to home and being less dependent on hospital facilities.

Laboratory professionals have some concerns about the quality of the results produced by POCT; at the same time, they may also be worried about POCT encroaching on their territory. Clinicians share some of the anxieties about the quality of the results, and the potential impact on their patients. Healthcare managers and policymakers are concerned with ensuring quality, in line with their laboratory and clinical colleagues, but are also concerned about the cost of care, and that of its individual components. While recognizing these sectarian interests, we acknowledge that we are all also concerned about the care of the patient.

Healthcare is a complex business. In most countries it suffers from being organized in a classical vertical structure, invariably being more efficient within individual departments than as a whole. This problem is particularly acute in the area of financial management, as exemplified in many countries by the process of reimbursement and allocation of resources to areas such as laboratory medicine. As a consequence, there is little incentive to develop or invest in new technology or interventions beyond the obvious therapies that could directly and immediately benefit patient outcome.

In recent years, many have called for a different approach to the organization and management of healthcare, an approach that puts the patient, instead of the healthcare professional or the provider organization, at the center of the process of care. This trend parallels the business environment's more systems-based approach. This parallel is important because business management techniques are now being increasingly employed in the delivery of healthcare-to improve quality, to enhance efficiency, and to contain costs.

Considering healthcare trends led us to realize that, invariably, when we communicate with laboratory professionals, clinicians, and other caregivers about POCT, we tend to focus on the device's analytical performance and on its cost. Less frequently-and only recently-do we talk about the outcomes or benefits. Even less frequently do we talk about the "whole systems approach" and how we need to change the process of care in order to really deliver the benefits. POCT is all about changing the process of care-delivering results quickly, so that a decision can be made and action taken. A consequence of this change of process will be an assessment of the value of the new process, and a reallocation of resources to make it happen. This is where it gets to be really challenging!

We are now beginning to see the term "outcomes" used by politicians, health policymakers, and healthcare managers. We have always focused on outcomes as part of our professional "mission statement," but have we really understood how we, each in our professional lives, influence health outcomes in the way we practice? POCT is not solely about analytical performance and cost, but about outcomes. However, in order to ensure improved outcomes, we must demonstrate how these can be delivered and how organizations must adapt as a result. Consequently, we must engage with managers and policymakers.

This book is therefore intended for healthcare managers and policymakers, to explain what POCT is, how it needs to be organized to ensure a reliable service, how the introduction of POCT will impact the way that healthcare is delivered and how the healthcare delivery system must change as a result, and what benefits are to be gained. The points made in the text are illustrated with examples taken from the peer-reviewed literature. We have drawn heavily from our earlier writings, particularly from a comprehensive review of the application of POCT in a wide range of settings.[1] We also appreciate how conversations with John Thompson on "business thinking" have informed this text.

Christopher P. Price

Andrew St John

[1]Price CP, St John A, Hicks JM, eds. Point-of-care testing, 2nd ed. Washington, DC: AACC Press, 2004:488pp.

Chapter 1

Point-of-Care Testing: What, Why, Where, and How?

The term "diagnostic test" is generally used to describe any form of investigation that assists the healthcare professional in making a decision about the care of an individual patient. The term "diagnosis" generally refers to any process of investigation that is associated with determining the presence or absence of a specific disease. The process of diagnosis involves observation—for example, of signs and symptoms—and the establishment of a working hypothesis against which information from a range of investigations (including simple observation) is used to test the hypothesis, hopefully to clearly support or refute the claim. If a hypothesis is refuted, then a new hypothesis is established and tested, perhaps with additional information being provided.

An alternative way of looking at this process and the parts played by investigations (i.e., diagnostic tests) is that the initial observation phase establishes a "pre-test probability" of having or not having a disease, and the investigations are used to establish a "post-test probability" of having or not having a specific disease. If the test has real value, it will have increased or decreased the probability significantly, such that the clinician can be more confident of the diagnostic decision ("ruling in" or "ruling out" the disease, respectively) (1,2).

WHAT IS POCT?

The history of medicine shows that all diagnostic tests evolved from the point of care, and this is certainly true in the case of laboratory medicine. The earliest reports of tests being performed on urine show that these tests were performed at the bedside. These tests were initially based on using the senses to examine the urine, and then evolved to using very basic chemical tests. Later these tests moved away from the patient's bedside into some form of side room, and in effect, the laboratory was formed.

Clinical medicine has developed in a similar manner: patient care facilities became hospitals, and laboratories evolved—most probably within the hospital setting. In the years that followed, as the knowledge of diseases advanced and technological innovation increased across the breadth of healthcare, there has been a shift from the physician attending the patient to the reverse occurring (Figure 1-1). This reversal has in many ways centralized the technology that is such an important part of clinical medicine and patient care.

Centralization has also been due to consumer growth and demand for greater efficiencies of delivery. The evolution of the retail sector offers an instructive analogy. The retail sector, crucially dependent on the transportation infrastructure, requires the customer to visit the shop (more and more so, the shopping complex) rather than the reverse.

Centralization can lengthen the diagnostic decision-making process. This may not be acceptable, either to the patient or to the clinician. Furthermore, in certain situations, a clinical decision may have to be made quickly in order to implement some form of intervention to improve the patient's immediate symptoms and longer-term health outcome.

Point-of-care testing (POCT) has therefore evolved, being defined as "any test that is performed at the time at which the test result enables a decision to be made and an action taken that leads to an improved health outcome." There can be both physical and temporal dimensions to the use of the term "point" and, as will become clear later, it is the temporal dimension that is most important in terms of health outcome (3). Some of the other names given to this form of testing have included "bedside" (4); "near patient" (5); "physician's office" (6); "extra-laboratory" (7); "decentralized" (8); and "off-site," "ancillary," and "alternative site" (9) testing. Several of these names allude to the origin of the request, namely the bedside, or to the use of the laboratory as the common modality of

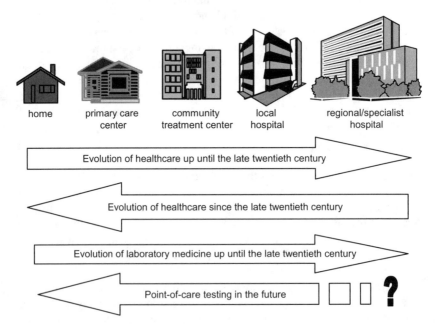

FIGURE 1-1. The evolution of healthcare from the bedside to the hospital setting up until the late 20th century, and the reversal that is occurring in the 21st century, together with the parallel evolution in laboratory medicine.

test delivery, and maybe hint at the limitations of this approach. Thus, physician's office testing evolved to meet the needs of the clinician wishing to make decisions at the time of the consultation—but also to retain some independence from the central laboratory, perhaps in terms of immediacy of delivery as well as financial independence. The use of alternative site testing sought to delineate testing that remained under the jurisdiction of the hospital but outside the traditional laboratory setting (10). However "alternative site testing" is a term that has also been used specifically in relation to blood glucose testing on parts of the body other than the conventional "finger stick" (11).

WHY CONSIDER POINT-OF-CARE TESTING?

A diagnostic test can be used to answer a clinical question, and the purpose of using a point-of-care test is to provide the answer more quickly. These are some typical clinical questions to which POCT can provide answers:

- Can I "rule in," or "rule out," diagnosis X now?
- Can I assess prognosis now?
- Can I initiate therapy now?
- What interventio n is best for this patient?
- What level of intervention should I use?
- Has my intervention been effective?
- Is this patient complying with the treatment prescribed?
- Will having this information on hand help in my consultation with this patient?

Note that these questions are generally related to an individual patient, and therefore provide a very patient-centered approach to care. Note also the potential impact of the ability to immediately provide a result. Some specific examples of questions that may be answered, or decisions that may be made following an obtained result, are given in Table 1-1. It is clear from this short list of questions that the part of the care process in which testing is included can be represented in a number of ways: for example, screening, diagnosis, prognosis, therapy selection, optimization, and compliance.

The underlying rationale for POCT is that providing the test result when the clinical question is asked will benefit the patient, the clinician or caregiver, and/or the health provider organization.

The diagnostic performance of any test will invariably have been established in a series of studies, primarily using a laboratory-based method. Hence the sole issue for POCT is whether the availability of the result within minutes of asking the clinical question will improve the health outcome for the patient, for the caregiver, for the healthcare provider organization, or possibly even for the purchaser. The health outcome from a management perspective can be differentiated into clinical, operational, and economic components and can be measured in a number of ways. This will be discussed in greater detail in a subsequent chapter, together with a number of illustrative case studies.

TABLE 1-1. Examples of the type of clinical questions relevant to the use of POCT

- Can urinalysis be used to "rule out" urinary tract infection in an asymptomatic at-risk woman?
- Can chlamydia screening reduce the prevalence of complications of infection?
- Does screening for urine protein give earlier detection of renal disease in at-risk populations?
- Does frequent self-monitoring of blood glucose improve glycemic control?
- Does self-monitoring of prothrombin time improve maintenance of a patient's INR?
- Does HbA1c at the time of the clinic visit improve compliance with treatment protocol?
- Does regular monitoring of serum cholesterol improve compliance with statin therapy?
- Does a rapid cardiac marker service enable effective "rule out" of a diagnosis of acute coronary syndrome in patients with chest pain?
- Does regular monitoring of blood gas and electrolytes reduce length of stay in intensive care?

WHERE IS POINT-OF-CARE TESTING ADVANTAGEOUS?

As noted earlier, the concept of POCT was born when the first tests were made on urine, and has evolved with both the discovery of new tests and testing technology as well as with changes in the way healthcare is provided. POCT can now be considered in a wide range of environments where a clinical decision is made (see Table 1-2). As individuals become empowered to take more responsibility for their own well-being, the definition of a clinical decision should be extended to include testing associated with "wellness" and health promotion/education.

TABLE 1-2. Settings in which POCT is performed

- Home
- Workplace
- Leisure facility
- Community pharmacy
- Health center (general practitioner/physician's office)
- Diagnostic and treatment center
- Outpatient clinic (physician's office)
- Ambulance/helicopter
- Mobile hospital
- Emergency room
- Admissions unit
- Operating room
- Intensive care unit (adult, pediatric, coronary care)
- Ward (unit)

MAKING THE DECISION TO USE POINT-OF-CARE TESTING

It is helpful to have an overview of the steps involved in generating results from diagnostic tests and in implementing the action that follows from the decision made upon receipt of results. This can help to

- understand the clinical need for the test;
- identify the logistics associated with test requisition, sample collection, and delivery;
- define the appropriate analytical method and its performance characteristics;
- identify the informatics challenges associated with delivery of the test result at the time it is needed;
- make the business case for the testing modality required; and
- develop the best framework for an audit of outcomes.

Figure 1-2 provides this overview, beginning with the patient consultation and ending with the implementation of the intervention that follows from the test result. It becomes immediately clear that from an operational standpoint, POCT offers a number of benefits:

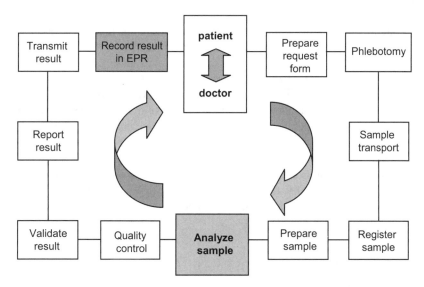

FIGURE 1-2. The steps involved in making a test request when using a laboratory service. Each step carries a risk, and so reducing steps reduces risk. The shaded boxes show the minimum number of steps for POCT.

- Reduction in the administrative work associated with test requesting.
- Avoidance or minimization of delay in a whole range of processes including
  - collecting the specimen according to the specified requirements,
  - transporting the specimen to a testing site,
  - registering receipt of the specimen,
  - processing specimens in a complex testing facility,
  - reporting the result and entering the result into the patient's record, and
  - getting the result to the caregiver making the original request.

Each of these steps represents a risk in the use of any diagnostic test, and so POCT, because it avoids or minimizes delay in these processes, reduces these risks. POCT therefore makes good business sense: "business thinking" on the optimal organization of any multicomponent process says that risk increases with the number of steps involved in the process; reducing the number of process steps reduces risk. Put another way, performing a test when it is requested reduces the risk of

- taking a specimen from the wrong patient,
- taking the wrong specimen,
- mislabeling the specimen,
- losing the specimen on the way to the laboratory,
- misidentifying the specimen upon receipt at the laboratory,
- performing the wrong investigations,
- reporting the wrong results,
- failing to transmit results back to the requestor,
- having the requestor fail to acknowledge the result and take any action, and
- delaying the time from request to decision and action.

On the other hand, there are also risks involved in using POCT. Foremost among these are concerns about the technical performance of the device that is used—both from the perspective of the quality of the testing device and the competence of the operator. There are also concerns about the quality of the sample, and about the reporting, interpretation, and archiving of the result.

Guidelines are now widely available on how to implement POCT in a way that attempts to minimize these problems (12–14). However, the most challenging task concerned with implementing POCT today is probably ensuring that changes in clinical practice occur, to make the best use of the rapidity with which the result can be delivered.

TECHNOLOGY FOR POINT-OF-CARE TESTING

The remainder of this chapter describes the technology available for POCT. We begin with a brief description of the analytical procedure, followed by the basic requirements of an analytical device and a discussion on how advances in technology have contributed to POCT devices. We conclude with descriptions of some specific devices in each of three major categories:

(1) handheld devices, (2) handheld devices with meter reading, and (3) benchtop devices.

Analysis and Analytical Devices

There are two objectives in any analysis: (1) to determine whether a substance (the *analyte*) is present and (2) to determine how much of it is present. All analyses require a *sample,* typically blood or urine. In this chapter we will not consider any test system that looks at the skin, e.g., pigmentation, change in architecture, etc. Table 1-3 lists examples of analytes that are measured using POCT. If the sample is blood, an analysis can be performed on plasma, serum, or cells (red cells, white cells, platelets, etc.); however, when POCT is used, the analysis is rarely performed on serum as this requires the sample to clot, which constitutes a time delay, as well as another step with an inherent risk. All analyses require a given amount of sample, yielding varying degrees of accuracy depending on the analytical method; therefore, there must be a means of ensuring that the correct sample volume is used.

TABLE 1-3. Examples of analytes and clinical settings that can be addressed using POCT

Home
- Blood glucose

Primary care
- Urinalysis
- Pregnancy test
- Chlamydia

Long-term disease management (primary care, pharmacy)
- HbA1c
- Urine albumin creatinine ratio
- Cholesterol
- International normalized ratio (INR)
- Brain natriuretic peptide

Emergency department
- Cardiac markers
- D-dimer
- Salicylate, acetaminophen, alcohol
- Blood glucose

Operating room
- Ionized calcium
- Parathyroid hormone

Intensive care unit
- Blood gases and electrolytes
- Blood lactate

In a few cases the presence of the substance of interest can be determined by a property of the sample, e.g., color, viscosity, or electrical charge, but in most instances the substance is recognized and quantified by its reaction with other substances, called *reagents*. If the substance of interest is recognized by color, then the sample is usually diluted with a simple reagent containing, for example, buffer made up in water; a similar diluent (diluting agent) is used for more complex reagents, but it may include other constituents. In some cases the blood sample itself provides the diluent.

Recognition of the analyte occurs in a number of ways, in part depending on the nature of its structure. Recognition can be achieved by use of an enzyme that recognizes the analyte as a substrate or co-factor, by an antibody raised against that specific analyte, or by a unique DNA or RNA sequence that complements the sequence of the analyte. The inherent characteristic of the analyte (color, viscosity, electrical charge, etc.) or the reaction between the analyte and the reagents will change the property of the mixture, which is then measured using a *detector* system; this change is called the *signal change*. The change in signal could occur through transmitted light, fluorescence, luminescence, electrical conductivity, or voltage (electrochemical).

If the substance of interest is to be quantified, then the *signal change* produced with the sample of interest (i.e., the patient's sample) is compared with the signal change produced with a sample of known concentration, which is called the *calibrator*. Establishing the relationship between the signal and the concentration is called the *calibration*. Early POCT systems required first running a calibration sample, then running the patient's sample. Comparing the two signals enabled the amount of analyte in the patient's sample to be calculated. In most modern POCT systems the reagents are manufactured to a high degree of reproducibility, and in large numbers so that the whole batch can be calibrated in the factory; this calibration holds for a defined period of time—as long as the systems are stored properly. The combination of the means of capturing the sample, holding the reagents, and bringing the sample and reagents together in a piece of technology is often referred to as the *device*.

Basic Requirements of a POCT Analytical Device

The user of the POCT technology, or "device," is likely to be a person who has not had a lengthy training in analytical sciences. Developers, evaluators, and managers responsible for implementing POCT in a healthcare environment must thus ensure that all POCT devices meet the following requirements:

- The device must be simple to use, i.e., require a minimum of steps to produce a result.
- The device and any associated reagents must be robust in terms of transportation, storage, and use, including calibration.
- The device must produce results concordant with the central laboratory and consistent with the clinical need.
- The device must be safe in terms of storage, usage, and disposal.

The most robust devices are those that enable the majority of the steps involved in performing an analysis to be encapsulated in the device, thereby minimizing the operator steps required. If the steps are similar to those the operator uses in other walks of life, then that familiarity brings confidence in the operation of the device. For example, some devices include the use of touch pads and insertion ports similar to those used in ATM systems.

Technological Innovation and Integration

POCT devices developed as a result of advances in a number of enabling technologies such as detection methods (detector), recognition methods (recognition), fluid handling (fluidics), and signal reading (transduction).

Detection and recognition of analytes are considered the primary enabling technologies. Yet these technologies have themselves been facilitated by other technologies associated with complex fluidic handling due to advances in papers, membranes, plastic molding, micro-machining, surface treatment, moisture barriers, layered material, chromatography, and capillary and pressure systems. Thus modern devices have eliminated the need for multi-step operations such as wiping to remove sample at the "end" of the prescribed reaction period, mixing, pipetting, addition of secondary reagents or washes, and timing, all of which carry a risk in terms of producing a reliable result.

Another major advance has been the ability to produce devices that are capable of reading a range of signals at low cost, but that still allow accurate and precise measurement. These devices include processors, displays, touch screens, photocells, and memory, as electronic components and circuit boards advanced in design and fabrication. The introduction of larger-scale reproducible reagent manufacturing techniques—such as dry reagent formulation, thin film casting, screen and inkjet printing, coating and dosing techniques, surface treatment, cassette and channel molding, and fabrication—has also played a role in lowering costs. In addition, the innovative use of certain raw materials that enhance reagent stability has allowed the low-cost manufacture of electrode sensors, dry reagents, and film-based devices.

Data processing, such as the analytical calibration and quality control methods that are invisible to the lay user but that have facilitated performance improvements toward laboratory standards, has been a crucial enabling technology. This is a result of the large-scale reagent production noted above, plus input and output devices, calibration algorithms, bar-coded lot identification, built-in quality control systems, and bi-directional instrument communication.

Designing for Reliability

Design of new devices must take into account system performance requirements. These performance requirements must not only meet clinical requirements but also recognize the settings in which devices might be used. Features must be designed into the device to account for environmental factors that might impact results, minimize errors, assure quality, and meet the user requirements. Detector, fluidic, transducer, and enabler technologies play key roles in defining the design approach.

Minimizing errors

Preventable errors include errors occurring in the sample, component, opera- tor, and performance stages of an analysis. Sample errors can be preanalytical or interference related, while operator errors can be procedural or accidental. Component errors can be stability- or fail-related, while performance errors can manifest as inaccuracy or imprecision. Choice of the detection and fluidic technology is the first step toward reducing or preventing these errors; trans- ducers and enablers can offer solutions in cases of limitations.

Accurate clinical analysis depends on steps that happen even before the sam- ple is placed in the analytical device. For example, there are guidelines for how specimens should be collected, handled, and stored prior to testing. Such criteria, originally defined for laboratory analyses, will generally also apply to POCT devices. The choice of the detection methods can reduce sensitivity to sample storage; for example, an immunoassay might use an antibody not dependent on an unstable protein conformation to work (15). In general, all devices are manu- factured to give stable results for specimens stored over several hours at room temperature, days for refrigeration, and months for freezing.

Unstable analytes can present problems even for a given detection method. A minimum response time can be required for a fresh sample. However, a "fresh time" can be hours in the case of an organism, minutes in the case of blood gases, or seconds in the case of a finger stick. As a general rule, POCT can obviate the concerns around sample stability, but stability issues should never be forgotten—for example, when a POCT device fails and the sample is taken to another testing site or sent to the laboratory.

Choice of fluidics can increase the "fresh time" window by acting as a col- lection, purification, and storage device. Examples include extraction solutions used for infectious disease tests such as *Helicobacter pylori*, influenza A/B, infectious mononucleosis, and streptococcus A (16). In whole blood testing, often a separation of plasma from cells eliminates irrelevant signals and insta- bilities from cellular material such as hemoglobin (17). One fluidic example is the use of a trapping layer placed directly over the signal-generating layer.

In general every effort is made to ensure that a device will offer complete or sufficient resistance to interference, but it is not always possible to achieve this due to the detection method or fluidics used. Several methods are used to obtain a reasonable protection from common biological components, deter- gents, preservatives, matrix effects, and drugs. Interference can be destroyed by reactions in the detector, as in the case of urinalysis and glucose testing sys- tems (18,19). Treatment with a diluent is also a common solution to this prob- lem (20–22). While most potential interferences are known and mentioned in the labeling, contamination of specimens with any additive may affect results and should be avoided.

Operator error can be procedural or accidental. Procedural errors increase if the device is technique-dependent; engineering more steps into the device reduces this area of risk. The placement of specimen into the reactor matrix, and the reactor matrix into the transducer, are key ergonomic factors. The use of calibration numbers, calibration solutions, sample treatment steps, and multi-step instructions are all factors in whether a device is prone to procedural errors. A major potential error exists when a pipette is required to measure an exact volume of sample, and this must be mitigated by stressing the importance of accuracy during training (23). Design can minimize errors by using detection and fluidic technology that requires the user to perform only one step, the other steps being "hidden" in the device. This is evident in the transformation of glucose systems that are now able to produce results directly from a finger stick, with autocalibration and with elimination of precise volume measurement (24). More recently, blood glucose strips that do not require calibrating the meter according to the batch of test strips has been developed; this has overcome another variation in performance (25).

Accidental errors are often not obvious to the laboratory professional but occur with untrained lay users, many of whom are not familiar with electronics and who lack understanding of analysis and clinical results. Error trapping and alarm reporting help minimize the impact of these errors, but built-in tolerances are often needed as repeat error messages reduce the devices' usefulness. Tolerances reflect a built-in safety measure. For example, an immunochromatography cassette minimizes the impact on sample handling by allowing overfills and producing error flags for underfills (22).

Component errors due to instability or failure must be prevented in the design stage. Stability of the detector and fluidics must be accounted for over the shelf life of the device and in the defined storage condition requirement. This can be accomplished either by minimizing any change in the consumables or by correcting for the change. Instability errors are typically invisible to the lay user until the product reaches a critical environment or age. Many ingredients and materials are used to solve instability problems (26,27).

Failure errors are prevented by simplicity in instructions and operation. Specimen application and placement of the disposable into the analyzer must be operator-independent. Read-times are typically set for convenience but not all detectors are read-time independent. Specimen application must prevent crossover and carryover between tests and leaching of reagent out of the device.

Factory calibration procedures can reduce accuracy problems. At the time of manufacture, calibration solutions with known concentrations of each analyte are introduced to the sensors, and the signals are then stored as set-points. When a patient sample is analyzed, the analyzer compares the signal obtained from the calibration solution to the signal from the patient's sample and calculates the analyte concentration. Re-usable sensors can use live calibration prior to testing; disposable sensors use representatives from one batch to characterize the lot. Most handheld devices must produce a numerical result with some degree of accuracy and minimal bias affecting the clinical decision limit.

Bias is inherent in the detection method and specimen variation or matrix effects (e.g., hematocrit effects). Dilution steps eliminate matrix effects in all cases (20). The degree of accuracy as defined by the range around the average can also vary with fluidic and detection method. A device can be pre-warmed to increase accuracy by proceeding with testing at a set temperature.

Assuring quality

Quality defines the essential performance requirements, and compliance with the protocol assures standardization. This is accomplished by three methods: (1) control systems, (2) checks, and (3) compliance. These methods ensure that "requirements" are met when the device reaches the market. When new devices are developed, the developers apply some form of these methods to each device, depending on the design. Technology is used to overcome the constraints that may be recognized in a system, or in the sample. As an example, the results produced by the early blood glucose testing systems were influenced by the hematocrit of the sample—a problem that does not exist in the majority of the more modern devices.

Control systems are used to ensure that the instrument and reagents meet the stated specifications for assurance of claims. Release of meters and reagents is performed with factory-designated calibration reagents. In-lot calibration control system characteristics defined by the manufacturing lot are entered into the instrument by assignment numbers. The transducer identifies the test to be performed by reading the code (electronic signature, bar code) on each disposable.

Checks are examples of safety controls that are now included in the device. This assures that the device has not failed in its the ability to store quality control data, information relating to date and time of operation, and operator and specimen identification. Checks are part of the trend away from subjective visual color reading to instrument reading of the reaction, to counteract the effects of color-blindness and differences in reading method and interpretation of color. Remote control checks of devices outside the central clinical laboratory are a new trend. Other systems now "lock out" the user every 24 hours unless a quality control function is completed. Control samples, both real or electronically simulated, are used as checks in a given time period. Immunochromatography strips are a good example of an application of checks for false results from the detector (28). Negative control lines are used to ensure that the reagent is present, while a positive control line confirms that the reagent is reactive to analyte. Fluidic control lines are used to verify correct flow through the device.

Compliance to standards allows greater quality of results and safety across manufacturers. Standards for safety, performance, medical decision limits, clinical and laboratory practice, and others are considered, along with processes such as risk analysis, traceability, and good development practices; together they assure that a certain level of work was applied before approval was granted. Standards for electrical discharge (CE), dangerous reagents, sampling device or lancet, and environmentally friendly disposal of the analytical consumables are accounted for in device designs.

User requirements

Convenience of use is a fundamental user requirement. One measure of convenience is the minimization of training or periodic retraining. Others include avoiding multiple steps for activation together with seamless procedures for calibration, quality control, data entry, sample presentation, reading of results, data transfer, and maintenance.

Durability is another absolute requirement. Durability is "built in" by choosing designs that accommodate display screen stability, drop tests, reliability tests, and "accelerated use" testing. When costs for the instrument or the disposable components are high, the device cannot effectively compete with the laboratory-based service. Physical size and weight requirements are important, as working space for the device is always limited.

User requirements depend on what technology can achieve, what the user generally accepts, what is "best in class," or what is possible within a generation. Today, glucose meters are as small as the smallest PDA, while the majority of blood gas meters are still benchtop size, most probably due to the complexity of the technology and the fact that they are generally reusable systems that measure a relatively large number of specimens each day. In some clinical settings a handheld device is ideal, whereas in others a benchtop system may be more useful (e.g., a multianalyte device in the emergency room or primary care physician's office). Another effort to meet user requirements is to keep "time to result" at a minimum. In cases where there is a requirement for high sensitivity, accuracy must not be compromised at the expense of size and timing, such as in the case of critical care devices (29–31). Data entry, management, and transfer are increasingly important requirements that went from "want" to "must-have" as technology allowed these features to be easily incorporated into devices.

Specific Types of Devices for Point-of-Care Testing

Most of the methods employed in POCT devices are based on those used in the laboratory. These methods are encapsulated into three broad categories:

1. Handheld devices
2. Handheld devices with meter reading
3. Benchtop devices

These devices may be used to measure single analytes or combinations of analytes, where the combination has some clinical relevance such as in urinalysis or cardiac markers. It is more likely that handheld devices are designed for single analytes, and that meters and benchtop systems may be capable of measuring more than one analyte—and often unrelated analytes. The potential advantage of this approach is that the operator has to be trained in only one operating system; having to learn how to use a number of different operating systems carries with it a risk of making errors. The examples discussed below illustrate the principles underpinning POCT device design. The coverage is not comprehensive, however, and the reader is thus referred to larger publications to appreciate the full spectrum of available analytical systems (32–34).

Handheld devices

Three major types of handheld devices are illustrated in Figure 1-3. In the first, probably considered to be the simplest, reagents are included in a porous matrix that is attached to some form of holder to facilitate easy handling. Sample is applied to the matrix, and it mixes with dry reagent contained in the matrix to produce a signal that is interpreted visually by the operator or read using a meter reader, which can provide a digital output (26,27). The meters can vary in size, from being either portable and handheld to larger and more static tabletop systems. In addition, systems can vary by the number of results produced in a fixed period of time.

In the second type of handheld device, reagents are held in a reaction cassette, into which the sample is inserted (captured) and then mixed with reagents in one or more processes. In this configuration the reaction signal is invariably interpreted with the aid of a meter reader. In a small number of cases the reading element of the device is built into the cartridge.

In the third type, which at present is the least common, reagents and specimen react on the surface of the device, and the signal is read on the surface, or coupled to some form of transducer that converts the signal to produce some form of electrical response. This type of system does not always need fluidic manipulations, but the reaction signal is invariably interpreted by a meter reader.

The most common commercial handheld device is a dipstick, most often used to measure various parameters in urine but also used for blood tests (35). As technology has advanced, strips that were once read with the human eye are now combined with meters for more accurate quantitation. Thus glucose strips also fall into the category of a simple handheld device, but since they are now almost always used in conjunction with a meter, they will be considered in more detail below. Despite the simplicity of dipsticks, critical operator factors include (i) the need to cover the whole pad with the sample and (ii) because the reactions often do not proceed to completion, the need to time the period between placing the sample on the pad and comparing the resulting color to a color chart. Developments of these single stick devices include two pads for measurement of different concentrations of the same analyte such as albumin (35), measurement of both albumin and creatinine to provide an albumin-creatinine ratio (35), and semi-quantitative measurement of up to 10 different urine analytes using reflectance technology (36).

Handheld immunosensors in various formats are also widely used to measure analytes such as HCG, cardiac markers, and infectious disease agents. In such devices the recognition agent is an antibody that binds to the analyte. Detection of the binding event or signal transduction occurs by either reflectance or fluorescence spectrophotometry. The devices can be used in a flow-through format where a heterogeneous immunoassay takes place in a

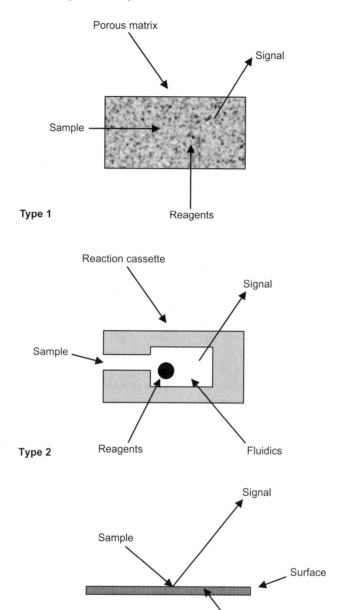

FIGURE 1-3. Diagram showing the major types of handheld devices used for POCT.

porous matrix cell containing a label of gold sol particles or colored latex that can be directly visualized (37). An important feature of this type of technology is the incorporation of a built-in quality control that indicates "positive" if all the reagents have been stored and the device operated correctly.

Lateral-flow chromatography formats are more widely available for detection of infectious disease agents such as chlamydia (38) and for cardiac markers such as CK, myoglobin, and troponin (28). These tests commonly use a lateral flow format in conjunction with analytical techniques such as enzyme immunoassay and immunochromatography with urine, serum, plasma, or whole blood samples. An alternative approach to immunoassay and immunochromatography involves light reflection and thin film amplification in what are termed optical immunoassays manufactured by Thermo Electron. The presence of an infectious disease antigen such as Strep A is detected through binding to an antibody coated on a test surface. Light reflected through the antibody film alone produces a gold background that changes to purple when the thickness of the film increases due to the presence of an antigen (Figure 1-4). The tests include built-in controls and provide results comparable to those provided by conventional microbiological assays but much more rapidly (39).

Handheld devices with meter reading

The addition of the meter adds a number of benefits to the core of the handheld device. Primarily, it enables quantitation of the signal produced. Thus it is possible to read the change in color or fluorescence using reflectance or transmittance photometry, or the change in electrochemical signal. Both of these approaches are epitomized in the blood glucose strip devices. In addition to enabling quantitation of the signal, the use of a meter reader also reduces operator variability, which may be due to interpersonal variation in visual acuity and degree of color blindness.

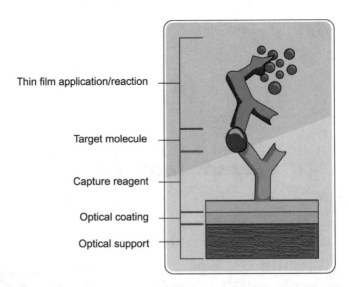

Thin film application/reaction

Target molecule

Capture reagent

Optical coating

Optical support

FIGURE 1-4.   Schematic diagram of the principles of an optical immunoassay (OIA) using thin film detection. *Courtesy of Thermo Electron Corporation, Waltham, MA, USA.*

One of the earliest examples of a handheld device using a reagent strip but linked with a benchtop meter is the Reflotron® system (Roche Diagnostics, Mannheim, Germany) (19). In this device multiple reagents are located adjacent to a plasma-separating layer in which a glass filter traps red blood cells, after which the plasma flows into a glass fiber reservoir. Initiation of the reaction takes place when the strip is inserted into a photometer that presses the reagent layers into contact with the plasma in the reservoir. The device offers a fairly extensive range of test strips, each recognized by the meter using a magnetic strip reader, which also contains the calibration data.

Devices for measuring glucose form the largest segment of POCT technologies. Since their introduction more than 25 years ago, glucose meters have undergone a steady stream of innovation and development, with the goal of making the devices smaller and easier to use with less risk of error, and of reducing potential interferences. Now many millions of diabetics around the world use this technology to monitor their blood glucose level. These devices are biosensors using one of three enzyme systems with photometric (reflectance) or electrochemical detection (24,25). The glucose strips are manufactured using thick-film technology, where the film is composed of several layers that have specific functions, including exclusion of erythrocytes, spreading of the sample to ensure homogeneous distribution, and a support layer that may also have reflective properties. These days most strips use electrochemical technology. Its advantage is that smaller meters can be used and the strips do not require wiping. A typical glucose meter is shown in Figure 1-5.

The I-stat system is an example of a handheld device that uses a cassette. The sensors or electrodes in this blood gas device are of a thin-film design with wafers constructed from metal oxides, using techniques similar to those used for computer chips (40). One of the advantages of using such a thin film is that the sensors can rapidly equilibrate with the blood sample, and there are no major delays between removing the sensor cassette from its packaging and adding the blood sample.

Benchtop devices

Benchtop devices are larger, offer more analytes, and can sometimes handle a number of samples. They have more sophisticated operator interfaces, including LCD display screens, touch screens, keypads, barcode readers, and printers. The latter are now almost obligatory in terms of ensuring that a hardcopy of the results is available. Patient safety and the need for positive patient, sample, and operator identification have led to incorporating bar code readers in many devices. These readers are also included for identifying reagents and other consumables to the system, some of which will incorporate factory calibration, as well as for informing the instrument how to process a particular test.

Control and communications systems are a major component of all benchtop POCT devices. Even the smallest device has a control subsystem that coordinates all the other systems and ensures that all the required processes for an

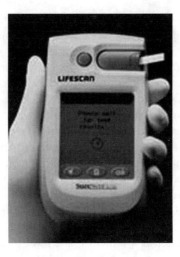

FIGURE 1-5. A home-use blood glucose meter. *Courtesy of Lifescan Inc., Milpitas, CA, USA.*

analysis take place in the correct order. Operations requiring control include temperature, sample injection or aspiration, sample metering, sample detection, mixing, incubation, timing of detection process, and waste removal. Fluid movement can often be accomplished by mechanical means, through pumps.

Blood gas instruments represent one of the earliest forms of POCT device. While instruments that measure just hydrogen ion (pH) and blood gases are still produced, many laboratories and critical care units now purchase devices that measure many additional parameters and are more correctly called critical care analyzers. The menu of tests includes blood gases; electrolytes; metabolites such as glucose, lactate, urea, and creatinine; co-oximetry; and bilirubin. In such a multi-analyte device the flow of blood, calibrants, wash, and quality control (QC) solutions through the device will be controlled by an extensive arrangement of detectors, valves, pumps, and software.

The device will use a range of analytical technologies including microelectrodes and thick-film technology for measurement of parameters such as glucose, lactate, and urea (41). These are biosensors using enzymes in reactions similar to those used in glucose strips, but the challenge for manufacturers has been to devise reusable enzyme-based sensors that will function after repeated exposure to whole blood samples. Use of this technology has enabled sensors to be incorporated with reagents and calibrators into a single cartridge or pack that is then placed in the body of a small- to medium-sized, portable critical care analyzer. Each pack measures a certain number of samples during a certain time period, after which it is a relatively simple procedure to replace. This type of technology (see Figure 1-6) is now available from a number of manufacturers (42–44), and its convenience means that it likely to be the format for all future critical care analyzers.

Many benchtop critical care instruments can also determine hemoglobin and its species, hematocrit, and bilirubin. CO-oximeters are an integral part of many

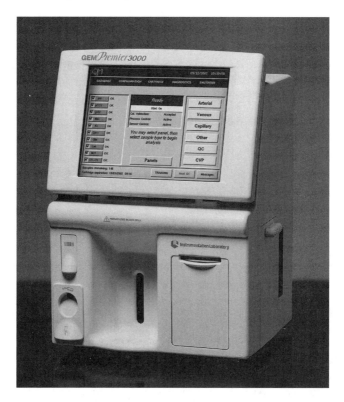

FIGURE 1-6. The Gem Premier cartridge-based critical care analyzer. *Courtesy of Instrumentation Laboratory, Lexington, MA, USA.*

benchtop analyzers and rely upon multiwavelength spectrophotometry, where light absorption by hemolyzed blood is measured at up to 60 or more wavelengths to determine the concentration of the five required hemoglobin species (45).

During several decades of development, these devices have not only been constructed to measure more analytes but also have gone through major advances in design and ease of use. Thus to overcome the expense and inconvenience of gas cylinders, several manufacturers use either liquid calibration systems, room air, or an internal O_2-zero solution, independently or in combination (46). Other features of these analyzers that contribute to their ease of use and minimize the risk of errors are shown in Table 1-4.

The basic analytical principles of many benchtop POCT chemistry and immunochemistry analyzers are similar to their counterparts in the main laboratory. An obvious key difference is that in order to minimize the sample handling involved with POCT devices, it is necessary to include either analytical methods or sensors that can directly utilize whole blood or include steps that remove the red cells prior to analysis. The devices described below all use whole blood samples, but in most cases serum or plasma can also be used.

Several analytical devices overcome the problem of interfering red cells by incorporating centrifugation within the instrument. One of these is the Piccolo®

TABLE 1-4. Features of critical care analyzers that minimize risk of errors and contribute to ease-of-use

- Long-life, maintenance-free electrodes or disposable sensor packs
- Touch screens as the user interface
- Software that can demand user and patient identification
- Built-in bar code scanners
- Sample aspiration instead of injection
- Reduced sample sizes
- Clot detection within analysis chamber
- Sample detection to prevent short samples
- Liquid calibration systems instead of no gas bottles
- Automated calibrations
- Automated quality control sampling
- Sophisticated QC programs including interpretation of data
- Connectivity to information systems allowing remote monitoring and control
- Built-in videos for training purposes

analyzer (Abaxis, Union City CA, USA) which incorporates a centrifuge not only to separate cells from plasma but also to distribute and mix sample and reagents (47). Other devices that can perform general chemistry tests include the Careside analyzer® (Careside, Culver City, CA, USA), which is based on single use test cartridges that includes a plasma separation step (48), and the LDX™ system (Cholestech, Hayward CA, USA), which has cartridge-based tests for lipids, glucose, and ALT. Both devices are designed to minimize the tasks required of the operator once a blood sample has been placed in the cartridge or device. The relative simplicity of the Cholestech device is indicated by the fact that it is classified as a CLIA-waived device (49).

Several devices are available to measure cardiac markers using lateral flow strips in conjunction with a reader. The Stratus CS® analyzer (Dade Behring, Wilmington, DE, USA) uses enhanced radial particle immunoassay to detect cardiac markers, including troponin (50). The device incorporates a centrifuge, which makes it one of the larger benchtop devices, but it also includes sampling direct from the blood collection tube, thus minimizing pre-analytical errors. This feature, together with its better analytical sensitivity for troponin compared to many other methods, makes it a popular POCT device.

Diabetes testing and monitoring is another clinical area where POCT can deliver patient benefits, and several small cassette or strip-based devices are available for measurement of HbA1c and for urine albumin. The Bayer DCA™ (Bayer HealthCare, Tarrytown, NY, USA) is a small benchtop device that can measure these key analytes as well as creatinine so that the albumin measurement can be corrected for urine concentration (51). This cartridge-based system uses a light-scattering immunoassay to measure glycated hemoglobin, together with a colorimetric assay for total hemoglobin. The cartridge is a relatively complex structure that contains antigen-coated latex particles, antibodies to HbA1c, and lysing reagents that are mixed following addition of the whole blood or urine sample. Measurement occurs when the cartridge is placed into a temperature-controlled reader (Figure 1-7).

(a)

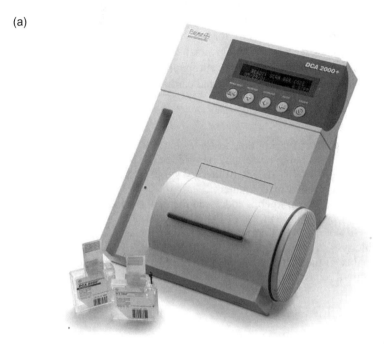

(b)

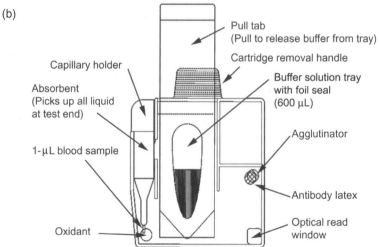

FIGURE 1-7. (a) The Bayer DCA 2000 HbA1C immunoassay instrument. *Courtesy of Bayer Healthcare, Tarrytown, NY, USA;* (b) schematic diagram of the cartridge. *Courtesy of Pugia MJ, Price CP. Technology of handheld devices for point-of-care testing. In: Price CP, St John A, Hicks JM, eds. Point-of-care testing. Washington, DC: AACC Press, 2004.*

In recent years several benchtop immunoassay instruments have appeared. These instruments potentially have a larger menu of tests available than those capable of being performed on the smaller strip-based tests, and yet are of a size that can be accommodated in clinics or doctors' practices. The Innotrac Aio!™ (Innotrac Diagnostics, Oy, Finland), for instance, is a fully automated random-access immunoanalyzer based on a universal all-in-one (AIO) reagent concept that uses time-resolved fluorescence of europium lanthanide chelates for detection of parameters such as CRP, HCG, and cardiac markers (52).

SUMMARY

- POCT was the first form of diagnostic testing, the earliest reports of which documented tests being performed on urine at the patient's bedside.
- Technology for POCT ranges from very simple "stick tests" to complex equipment controlled by computers.
- Technological innovation has addressed many of the initial concerns about the robustness and ease of use of POCT devices.
- The quality of the POCT service is ensured by the selection of appropriate technology.
- The quality of the POCT service is ensured by competent operators working within a robust clinical governance framework.

REFERENCES

1. Sackett DL, Haynes RB, Guyatt GH, Tugwell P. Clinical epidemiology: a basic science for clinical medicine, 2nd ed. Boston: Little, Brown, and Company, 1991:441pp.
2. Sackett DL, Strauss SE, Richardson WS, Rosenberg W, Haynes RB. Evidence-based medicine: how to practice and teach EBM, 2nd ed. Edinburgh: Churchill Livingstone, 2000:261pp.
3. Price CP. Point-of-care testing: potential for tracking disease management outcomes. Dis Manage Health Outcomes 2002;10:749–61.
4. Oliver G. On bedside urine testing. London, UK: HK Lewis, 1984:1–128.
5. Marks V, Alberti KGMM, eds. Clinical biochemistry nearer the patient II. London, UK: Churchill Livingstone: 1988:192pp.
6. Mass D. Consulting to physician office laboratories. In: Snyder JR, Wilkinson DS, eds. Management in laboratory medicine, 3rd ed. New York: Lippincott, 1998:443–50.
7. Price CP. Quality assurance of extra-laboratory analyses. In: Marks V, Alberti KGMM, eds. Clinical biochemistry nearer the patient II. London, UK: Bailliere Tindall, 1987:166–78.
8. Ashby JP, ed. The patient and decentralized testing. Lancaster, UK: MTP Press Ltd., 1988:128pp.
9. Handorf CR. College of American Pathologists conference XXVIII on alternate site testing: introduction. Pathol Lab Med 1995;119:867–71.
10. O'Leary D. Global view of how alternate site testing fits in with medical care. Arch Pathol Lab Med 1995;119:877–80.

11. Tamada JA, Garg S, Jovanovic L, Pitzer KR, Fermi S, Potts RO. Noninvasive glucose monitoring: comprehensive clinical results. Cygnus Research Team. JAMA 1999;282:1839–44.
12. Clinical and Laboratory Standards Institute. Point-of-care in vitro diagnostic (IVD) testing: approved guideline. CLSI Document AST2-A. Wayne, PA: CLSI, 1999.
13. England JM, Hyde K, Lewis SM, et al. Guidelines for near patient testing: haematology. Clin Lab haem 1995;17:300–9.
14. Freedman DB. Clinical governance: implications for point-of-care testing. Ann Clin Biochem 2002;39:421–3.
15. Spitznagel TM, Clark DS. Surface density and orientation effects on immobilized antibodies and fragments. Biotechnology 1993;11:825–9.
16. Campos J. Diagnosis of infectious disease with point-of-care assays. J Clin Lig Assay 2002;25:333–41.
17. Kilpatrick ES, Rumley AG, Myint H, et al. The effect of variations in hematocrit, mean cell volume, and red blood cell count on reagent strip tests for glucose. Ann Clin Biochem 1993;30:485–7.
18. Pugia MJ. The technology behind diagnostic strips. Lab Med 2000;31:92–6.
19. Steinhausen RL, Price CP. Principles and practice of dry chemistry systems. In: Price CP, Alberti KGMM, eds. Recent advances in clinical biochemistry. Edinburgh: Churchill Livingstone, 1985:273–96.
20. Guthrie R, Hellman R, Kilo C, et al. A multisite physician's office laboratory evaluation of an immunological method for the measurement of HbA1c. Diabetes Care 1992;15:1494–8.
21. Buechler KF, Moi S, Noar B, et al. Simultaneous detection of seven drugs of abuse by the Triage™ panel for drugs of abuse. Clin Chem 1992;38:1678–84.
22. Apple FS, Christenson RH, Valdes R, et al. Simultaneous rapid measurement of whole blood myoglobin, creatine kinase MB, and cardiac troponin I by the Triage Cardiac Panel for detection of myocardial infarction. Clin Chem 1999;45:199–205.
23. James DR, Price CP. Reflotron assays for alanine aminotransferase and γ glutamyltransferase in whole blood samples evaluated. Clin Chem 1987;33:826–9.
24. Henning TP, Cunningham TP. Biosensors for personal diabetes management. In: Ramsay G, ed. Commercial biosensors. New York: John Wiley & Sons, 1998:3–46.
25. Baum JM, Monhaut NM, Parker DR, Price CP. Improving the quality of self-monitoring blood glucose measurement: a study in reducing calibration errors. Diabetes Technol Ther 2006 (forthcoming: accepted for publication).
26. Zipp A, Hornby WE. Solid phase chemistry: its principles and application in clinical analyses. Talanta 1984;31:863–77.
27. Walter B. Dry reagent chemistries. Anal Chem 1983;55:498A–514A.
28. Collinson PO, Gerhardt W, Katus HA, et al. Multicenter evaluation of an immunological rapid test for the detection of Troponin T in whole blood samples. Eur J Clin Chem Clin Biochem 1996;34:591–8.
29. Severinghaus JW. The invention and development of blood gas analysis apparatus. Anesthesiology 2002;97:253–6.
30. Hedlund KD, Oen S, LaFauce L, Sanford DM. Clinical experience with the Diametrics IRMA (Immediate Response Mobile Analysis) blood analysis system. Perfusion 1997;12:27–30.

31. Kozlowski-Templin R. Blood gas analyzers. Respir Care Clin N Am. 1995; 1:35–46.
32. Price CP, St John A, Hicks JM, eds. Point-of-care testing, 2nd ed. Washington, DC: AACC Press, 2004;488pp.
33. Nichols JH, ed. Point-of-care testing: performance improvements and evidence-based outcomes. New York: Marcel Dekker Inc., 2003:500pp.
34. Kost GJ, ed. Principles and practice of point-of-care testing. Philadelphia: Lippincott Williams and Wilkins, 2002:654pp.
35. Pugia MJ, Lott JA, Clark LW, et al. Comparisons of urine dipsticks with quantitative methods for microalbuminuria. Eur J Clin Chem Clin Biochem 1997; 35:693–700.
36. Lott JA, Johnson WR, Luke KE. Evaluation of an automated urine chemistry reagent-strip analyzer. J Clin Lab Anal 1995;9:212–7.
37. Valkirs GE, Barton R. Immunoconcentration™. A new format for solid-phase immunoassays. Clin Chem 1985;31:1427–31.
38. Chernesky M, Jang D, Krepel J, et al. Impact of reference standard sensitivity on accuracy of rapid antigen detection assays and a leukocyte esterase dipstick for diagnosis of chlamydia trachomatis infection in first-void urine specimens from men. J Clin Microbiol 1999;37:2777–80.
39. Fries SM. Diagnosis of group A streptococcal pharyngitis in a private clinic: comparative evaluation of an optical immunoassay method and culture. J Pediatr 1995;126:933–6.
40. Erickson KA, Wilding P. Evaluation of a novel point-of-care system: The iStat portable clinical analyzer. Clin Chem 1993;39:283–7.
41. D'Orazio P, Maley T, McCaffrey RR, et al. Planar (bio)Sensors for critical care diagnostics. Clin Chem 1997;43:1804–5.
42. Jacobs E, Nowakowski M, Colman N. Performance of Gem Premier blood gas/electrolyte analyzer evaluated. Clin Chem 1993;39:189–93.
43. Lindemans J, Hoefkens P, van Kessel AL, et al. Portable blood gas and electrolyte analyzer evaluated in a multi-institutional study. Clin Chem 1999;45:111–7.
44. Magny E, Renard MF, Launay JM. Analytical evaluation of Rapidpoint 400 blood gas analyser. Ann Biol Clin (Paris). 2001;59:622–8.
45. Brunelle JA, Degtiarov AM, Moran RF, Race LA. Simultaneous measurement of total hemoglobin and its derivatives in blood using CO-oximeters; analytical principles; their application in selecting analytical wavelengths and reference methods; a comparison of the results and the choices made. Scand J Clin Lab Invest Suppl 1996;224:47–69.
46. Mollard J-F. Single phase calibration for blood gas and electrolyte analysis. In: D'Orazio P, ed. Preparing for critical care analyses in the 21st century. Proceedings of the 16th International Symposium. Washington, DC: AACC Press, 1996.
47. Boncheva M, Pascaleva I, Dineva D. Performance of POCT-chemistry analyzer Picollo (Abaxis) in primary health care. General Medicine 2002;4:28–31.
48. Schultz SG, Holen JT, Donohue JP, Francoeur TA. Two-dimensional centrifugation for desktop clinical chemistry. Clin Chem 1985;31:1457–63.
49. Cobbaert C, Boerma GJ, Lindemans J. Evaluation of the Cholestech LDX desktop analyser for cholesterol, HDL-cholesterol, and triglycerides in heparinised venous blood. Eur J Clin Chem & Clin Biochem 1994;32:391–4.
50. Kamm C, Elser R, Eitel D, Napier J. Evaluation of the Stratus CS® analyzer and three cardiac markers. Clin Chem 1998;44:1457–63.

51. Pope RM, Apps JM, Page MD, et al. A novel device for the rapid in-clinic measurement of hemoglobin A1c. Diabet Med 1993;3:260–3.
52. Hedberg P, Valkama J, Puukka M. Analytical performance of time-resolved fluorometry-based Innotrac Aio! cardiac marker immunoassays. Scand J Clin Lab Invest 2003;63:55–64.

Chapter 2

Organization and Management of Point-of-Care Testing

This chapter outlines the management processes that are required to implement clinically effective and safe point-of-care testing (POCT). It is assumed that the need for POCT has been established and an appropriate business case has been prepared to support POCT. These processes then need to be considered:

- Establishing a POCT Coordinating Committee and POCT Policy
- Considering accreditation and clinical governance requirements
- Instituting POCT Guidelines
- Procuring and evaluating equipment
- Establishing quality assurance procedures
- Organizing maintenance and inventory control
- Preparing documentation
- Training and certifying staff
- Conducting audits

The chapter concludes with a brief review of the impact of information technology on the practice of POCT. In addition, for those who have not yet established POCT in their institutions, but who wish to do so, the steps required to prepare a business case for POCT are provided.

POCT COORDINATING COMMITTEE AND POCT POLICY

Historically POCT has often developed within institutions on an ad hoc basis, i.e., without consulting the healthcare administration, and particularly without involving the laboratory. This occurred for a number of reasons, often because clinicians believed, sometimes justifiably, that laboratories were not providing required levels of service. As a result, clinicians organized their own diagnostic testing. In addition, clinical staff were, and still are, seduced by the marketing claims of some POCT device manufacturers, claims suggesting that all devices are simple and easy to manage. While POCT can and does occur successfully without involving laboratories, the collective experience throughout the world is that there are benefits to be gained through making POCT part of the overall diagnostic medicine service.

This change in thinking is due to a number of factors. First is the realization that not all POCT devices are simple to use, and that they often need to be operated by trained laboratory professionals to achieve optimum performance. Second is the increasing demand that all testing processes, including POCT, incorporate quality management systems (this demand has increased with the introduction of clinical governance) (1–3). Third, the laboratory will usually have to provide backup when the POCT service fails, for whatever reason. Lastly, within any institution, there must be a strategic approach to all diagnostic services to ensure that all testing is clinically and cost effective. All of the above can only be achieved by a management structure that includes all the relevant parties or stakeholders, including the laboratory.

Thus many healthcare providers have appointed a POCT Coordinating Committee to oversee all POCT within their institutions. The exact makeup of this Committee depends upon the institution, but it should include representatives of those who use and those who deliver the service; a patient representative might also be advisable. Since in the majority of cases the laboratory is providing the training, support, and backup for the POCT service, it is most common for a senior clinical scientist or pathologist to chair the POCT Committee. Ideally the POCT committee should report to the individual within the organization responsible for the quality of the clinical service, such as the Medical Director.

The POCT Coordinating Committee should draw up a POCT policy, the aim of which is to deliver a high quality service as well as provide a demonstrable means of protecting the interests of the patient. Without such a policy many problems are likely to emerge, not the least of which can be uncontrolled purchase and implementation of POCT (as has happened in the past). The POCT policy details all the major roles and responsibilities of a POCT Coordinating Committee (Figure 2-1).

A key member of the Committee is the person responsible for implementing POCT policy and delivering the service. Such people are often called POCT Coordinators or POCT Managers and, depending on the extent of the service, may require additional staff to carry out all of the necessary tasks associated with the service. At a local level, the Coordinator or Manager needs to define the details of the service that will be provided (this is sometimes called a Service-Level Agreement [SLA]) (4). Such agreements are useful in defining the expectations and responsibilities of all those who are involved in, and in receipt of, the POCT service. A typical SLA is shown in Table 2-1.

ACCREDITATION AND CLINICAL GOVERNANCE OF POCT

In most developed countries all laboratory services are subject to accreditation. In many but not all cases, the accreditation requirement has been

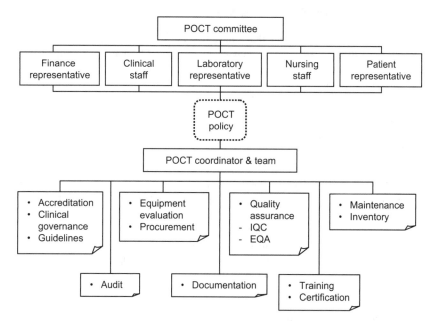

FIGURE 2-1. A typical management structure for the organization and management of point-of-care testing within a healthcare provider organization.

extended to POCT. Thus the Clinical Laboratory Improvement Amendments of 1988 (CLIA) legislation in the United States stipulates that all POCT must meet certain minimum standards (5,6). The Centers for Medicare and Medicaid Services, the Joint Commission on Accreditation of Healthcare Organizations, and the College of American Pathologists take on the responsibility for site inspection, and each is committed to ensuring compliance with testing regulations for POCT (7). These regulations govern how POCT should operate and thus they should be reflected both in the POCT policy and the overall management of POCT. Even in those countries where accreditation extends only to the central laboratory, the standards encompassed in these guidelines remain highly relevant to POCT and are worth consultation. Accreditation in many European countries is developing in accordance with ISO standard TC 212 (8).

Clinical governance is a recent development in healthcare management. It began in the UK but is being adopted in other developed countries (9). Whereas accreditation is primarily aimed at ensuring technical competence, clinical governance is about ensuring that POCT is clinically effective and minimizes the risk to patients. Thus clinical governance of POCT includes special attention to continuing professional development of all staff involved with POCT, and restricting POCT to those staff who have received recognized training, have demonstrated their competence to practice, and who participate in continuing professional development. Taking an evidence-based approach to clinical laboratory medicine is another key part of clinical governance, and

TABLE 2-1. A typical Service Level Agreement (SLA)

- Ownership of the equipment: this is most likely the clinical area in which it is sited, but some POCT systems may be owned by the central laboratory, or there may be a separate budget for POCT.
- Lines of responsibility for
  - results and how these will be reported and integrated into the patient record;
  - critical limits and action to be taken;
  - risk management;
  - health and safety, including infection control, disposal of blood and sharps, and maintaining the POCT equipment and surroundings in an uncontaminated condition;
  - equipment maintenance, if not by laboratory staff; and
  - management of regulatory compliance and accreditation of the POCT service (as required by the country or state concerned).
- Connectivity: patient data management, interface with the LIMS.
- Maintenance: staff time for maintenance and any consequent costs.
- Contact details for maintenance problems or troubleshooting.
- Operating costs: reagents and other consumables, including QC and EQA.
- Billing (if applicable).
- Stock control: storage and ordering arrangements.
- QC and EQA (proficiency testing):
  - who will perform;
  - procedures to be followed when these are out of range.
- Staff training and competency; agreement and registration of key trainers if appropriate.
- Arrangements for regular checks and reaccreditation.
- Test protocol, covering patient preparation, sample preparation, and reporting and interpreting results.
- Standard Operating Procedure for the POCT system:
  - analytical technique,
  - reference to QC and EQA procedures, and
  - recording and reporting results.
- Records: what records will be kept, for how long, and who will have responsibility and access.
- Audit: what audits will be conducted and the line of responsibility and processes for implementing any system improvements deemed necessary from audit results.

the audit process (see below) is used to ensure that the service is meeting the clinical purpose for which it is intended.

POCT GUIDELINES

Numerous POCT-related publications are available to help organizations implement POCT for the first time. In particular, various professional bodies and individuals have prepared guidelines for POCT that contain useful information that can be adapted to the needs of individual organizations. A list of these, which is not exhaustive, is shown in Table 2-2.

TABLE 2-2. Guidelines for POCT

Print, from Organizations

- Guidelines for Implementation of Near Patient Testing, 1993. Available from the Association of Clinical Biochemists, 130–132 Tooley St., London SE1 2TU, UK.
- International Standard ISO 22870(E). Point-of-Care Testing (POCT)— Requirements for Quality and Competence. Switzerland: International Organization for Standardization, 2006;pp 11.
- Point-of-Care In Vitro Diagnostic (IVD) Testing: Approved Guideline. CLSI Document AST2-A. Wayne, PA: CLSI, 1999.
- Point-of-Care Blood Glucose Testing in Acute and Chronic Care Facilities: Approved Guideline, 2nd ed. CLSI Document C30-A2. Wayne, PA: CLSI, 2003.
- Near-Patient Testing: A Statement of Best Practice for Scotland. Edinburgh: Scottish Office Department of Health, National Advisory Committee for Scientific Services, 1996.

Print, Citations

- Haeckel R et al . Good medical laboratory services: a proposal for definitions, concepts, and criteria. Clin Chem Lab Med 1998;36:399–403.
- Kost GJ. Guidelines for point-of-care testing. Improving patient outcomes. Am J Clin Pathol 1995;104 (Suppl 1):S111–27.

Web

- The Joint Working Group on Quality Assurance of Pathology (UK) (http://www.acb.org.uk/docex/Docs/NPT/8.pdf)
- Medicines and Healthcare products Regulatory Agency (UK) DB 2002(03); Management and Use of IVD Point of Care Test Devices (http://www.mhra.gov.uk)
- Guidelines for Perfusionists in Canada (http://www.perfusion.ca/categ/poct.html)
- German Working Group on Medical Laboratory Testing (http://www.dglm.de/~Ziems/poct_engl.htm)
- Introducing Point-of-Care Testing in General Practice; Department of Health & Aging, Australia (http://www.aacb.asn.au/pubs/poct_ps.pdf)
- Point-of-Care Testing: Position Statement, 2002; Australasian Association of Clinical Biochemists (http://www.aacb.asn.au/pubs/poct_ps.pdf)

Equipment Evaluation and Procurement

One of the first responsibilities of implementing a POCT service is to choose the device that is best suited to the task. There are many criteria for selecting analytical equipment, and some of these are shown in Table 2-3. Here we will focus on the criteria that are particularly important to POCT devices.

As mentioned earlier, POCT devices can be marketed as deceptively easy to use. It is true that manufacturers have greatly simplified the operator

TABLE 2-3. Evaluation criteria for POCT systems

- Compliance with regulatory standards.
- Connectivity: POCT1-A compliant for patient results transfer to LIMS and patient record, potential for remote monitoring and management of equipment from the laboratory; whether existing systems can be used.
- Bar code recognition for operators, reagents (consumables), patients.
- Analytes required (on separate analyzers or same—can be important in cases of small sample size).
- Methodology; comparability with existing methods (in central or decentralized testing).
- Any existing equipment standardization policy.
- Expected test number throughput; continuous or sporadic.
- Precision, accuracy, sensitivity, and specificity required, which should be appropriate for the clinical need.
- Specimen type and any special requirements or concerns about sample size or quality (e.g., neonates or fetal scalp specimens).
- Comparability with the laboratory result, particularly with different specimen types (e.g., blood or plasma glucose) and different analytical methods.
- Degree of portability required.
- Cost: capital purchase, leasing or "reagent rental."
- Running costs: maintenance contract, consumables, analytical quality materials, staff, software licenses and connectivity costs (including any software upgrades).
- Unauthorized operator lockout.
- QC:
  - AutoQC and regulatory compliance
  - QC lockout
- Availability of External Quality Assessment (proficiency testing) materials.
- Space available: type of facilities required for safe siting of equipment and any storage space needed (e.g., for consumables).
- Particular siting needs (e.g., temperature or humidity), to be matched with availability in clinical area.
- Maintenance requirements—clinical or laboratory staff time.
- Maintenance manuals supplied.
- Ease of use.
- Nature of any training material or sessions provided by the supplier (for clinical users or laboratory staff who will maintain the equipment).
- Warranty (period of free attention and repairs after purchase).
- Choice of service contracts, if appropriate.
- Supplier's engineer availability and response times.
- Supplier's help line.
- Any backup (e.g., replacement equipment on loan) available from the supplier when equipment is out of service for repairs.
- Trial period available for equipment in proposed site.
- Safety features (e.g., regarding infection risk and decontamination).

interface of many devices. Glucose meters provide perhaps the best example of how successive generations of instrumentation have steadily reduced the number of operator steps (10). However, manufacturers also acknowledge that no device is foolproof, and that equipment should be selected with that proviso in mind.

Therefore, a key aspect of selecting equipment is considering the user. The laboratory professional is not the typical user of POCT. Instead, the typical user group includes people such as clinicians, nurses, and other caregivers. These caregivers are unlikely to have had formal training in the use of analytical devices. To some degree, this lack of experience can be overcome with training, but training is unlikely to go beyond preliminary troubleshooting. Furthermore, people close to patients are busy and likely to face far more distractions than those people in a laboratory who are largely dedicated to operating a device or instrument.

Taking these considerations into account, then, the better POCT devices are those that require a minimum of operator steps. Ideally these should not be time- or sample-volume-dependent, nor should they require the operator's continual attention. A recent evaluation of several POCT devices shows that some devices do not fulfill these basic operator requirements; consequently major operator errors could occur (11).

Operator considerations are not the only criterion for choosing an appropriate POCT device. Other considerations relate to the analytical performance of the device. Any choice of device must consider the instrument's precision and accuracy, as well as the degree of agreement between results from the POCT device and results from the central laboratory. Ideally the desired precision and accuracy will be determined and agreed upon beforehand (setting so-called "quality specifications"), taking into consideration issues such as the biological variation in the analyte to be tested and the nature of the clinical decision that will be based on the result (12). Discussions with professional colleagues who are conducting POCT will reveal that this lack of agreement is a continuing problem, partly because the analytical technologies are often fundamentally different.

Once devices have been selected, it is essential to evaluate them before adopting them into routine practice. Yet time constraints often prevent an extensive review. One source for data about analytical performance is available from formal evaluations performed by a variety of different agencies or laboratory professionals and published in the peer-reviewed literature. Unfortunately a major problem with such data is that it is frequently obtained by staff skilled in the use of testing devices. Naturally, the analytical performance of the device in the hands of nurses or other staf who might be using the device would be much more relevant; data in the literature shows that the experience of "unskilled" vs. "skilled" users can frequently be different, and usually the experience of the typical POCT user is worse than that obtained by laboratory staff (13). Thus an in-house evaluation by those using the device is always preferable, but this evaluation should also include testing by laboratory personnel as a means to assess potential pitfalls, help with troubleshooting of later problems, and determine training routines. The basic evaluation of analytical performance has been addressed in a number of protocols, of which the Clinical and Laboratory Standards Institute [CLSI; formerly the National Committee for Clinical Laboratory Standards (NCCLS)] protocol is probably the most comprehensive and up-to-date (14).

Another aspect of device evaluation involves considering reagent requirements. For example, most POCT devices use pre-packed reagents, some of which require storage under special conditions (including refrigeration). Changes of different reagent batches may require extensive calibration procedures, all of which add to the complexity of the operation. Management needs to consider these requirements and how they might impact day-to-day operations—for example, will the busy nurse remember to remove the reagent packs from the refrigerator before testing? As mentioned before, if there is a choice of device, and other aspects being equal, the device that has a minimum of calibration, storage, and other logistical requirements might be the method to adopt.

Yet still another factor to consider when choosing equipment is the reality that no one supplier has the best devices for all analytes. Therefore, one must pick and choose, perhaps buying all glucose meters from one manufacturer and all blood gas devices from a different manufacturer. The administrative advantages of this are obvious, and there may be cost benefits as well, particularly when one buys "in bulk," such as with glucose meters. In this situation it is important to have good relationships with all POCT vendors. Countries such as the UK, incidentally, have taken steps to ensure good communication between suppliers and purchasing authorities (4). Regular communication between the laboratory staff and the suppliers of POCT devices can also help to avoid uncontrolled purchasing of devices by clinical staff without consulting those responsible for POCT services.

Examining POCT costs is a key part of the evaluation process. This examination should primarily focus on comparing the cost of different devices that perform the same test. The comparison should include not only the cost of equipment, reagents, and other consumables, but also of the labor required to operate the device. Simpler devices are not only less prone to errors but are also cheaper to operate in that they might require less labor. In addition, it's important to compare the POCT costs to those of the central laboratory. In relation to equipment, reagent, and labor costs, POCT will almost certainly be more expensive, but the clinical and economic benefits that might come from providing more rapid results must also be taken into account as part of the business case for POCT (dealt with later in this chapter).

Quality Assurance Procedures

The need to adopt comprehensive quality assurance processes for all POCT devices has been one of the major factors in persuading clinical staff and healthcare authorities that the laboratory is most suited to being responsible for POCT. Technological progress has simplified, and in some cases completely automated, the quality control (QC) procedures of certain POCT devices, and further progress will undoubtedly occur in this area. However, even with these technological advances, other quality assurance procedures still must be performed to achieve clinically and cost-effective testing.

For many POCT devices, operators must perform both Quality Control (QC) and External Quality Assurance (EQA) or Proficiency Testing procedures. QC is essentially about monitoring performance on a day-to-day basis while QA is a longer-term view and compares other but similar devices, often in different institutions. The details of how these procedures are conducted will not be described here and readers are referred to other more comprehensive texts (15,16). Instead, this section focuses on the technological advances in POCT device design, and how these are impacting quality assurance.

Technological Advances in Quality Control Management

With the exception of glucose testing, the most common types of POCT devices are those designed for measuring blood gases and other critical care parameters. Many countries have deployed these devices outside the laboratory for 15–20 years. Technological progress has been significant in all aspects of these devices, but despite miniaturization and longer life-time sensors or electrodes, the latter are still subject to drift and therefore need periodic calibration and monitoring with quality control samples. As a consequence these multiple-use devices are subject to what might be called traditional or statistical QC procedures, similar to those used on larger batch-type analyzers in the central laboratory.

The most common QC practice for critical care and other multiple-use analyzers involves measuring a liquid QC sample once per shift or every eight hours using three different levels of material over a 24-hour period. Sophisticated software now processes and plots the accumulated QC data, including algorithms such as Westgard rules, that allow users to determine how they will accept or reject QC data. Manufacturers of critical care systems have now introduced a degree of automation to QC processes. Some of these have been labeled Auto-QC and refer to processes similar to the automated calibration procedures that have been available on these devices for many years. A key part of the design is to ensure, as much as possible, that the Auto QC sample follows the same pathway and process as a patient's sample. Most systems allow the operator to set the time intervals at which the material is sampled, and most include software systems that monitor the results. If results are abnormal, the device can automatically repeat the analysis, and if it is still abnormal, then the software will shut down the relevant analyte channel or even the whole instrument (17).

The advantages of such systems are that QC samples are guaranteed to be analyzed at the specified times and if they fail, users can be prevented from accessing, or locked out of using, individual parameters or the complete instrument. Furthermore, given that these systems can be monitored and operated remotely, significant personnel resources can be saved, because remote instruments do not need to be visited so frequently by the people responsible for their management (18). While these packages allow for more frequent QC sampling and the use of algorithms such as Westgard rules, it is likely that most users are adopting the usual practices of three or two QCs per day and rejecting results greater than two or three standard deviations (SDs) from the mean. Although this complies with the current regulatory requirements in many countries, this standard routine has been criticized for only detecting major errors and allowing long periods (i.e., up to eight hours) to elapse before an error might be detected (19).

To address these criticisms, some manufacturers of critical care instruments have recently taken automated QC a stage further. One such system uses a multi-use, single cartridge containing all sensors and reagents that include so-called process control materials. These are akin to liquid quality control samples, except that they are analyzed much more frequently than two or three times a day; the frequency ranges from one every three minutes to one every 24 hours depending upon the analyte and the stability of the measurement system. Westgard rules-based software within the instrument monitors the data from all these measurements and makes any necessary adjustments to sensor drift and slope, applying corrective actions depending upon the degree of abnormality. In essence, while the process is not quite real-time QC, it does significantly shorten the period for error detection (20).

Along similar lines, another type of device incorporates a multilevel internal quality control system that includes electronic checks as well as two other internal QC systems. These all operate automatically and it is claimed that additional analysis of internal quality control materials is not required (21).

Single-use devices such as the glucose or cardiac marker systems also incorporate features that are designed to both improve and automate the quality control of the analytical process. The capabilities of modern electronics and miniaturization mean that it is possible to build into even the smallest analytical device a wide variety of electronic checks, and these have undoubtedly contributed to the overall improved performance of many POCT devices in the last decade. The concept has been used for many years in spectrophotometers and similar instruments through features such as photometer, filter, and general voltage checks. Thus, in POCT devices these electronic checks are frequently used, with the electrical signal check substituting for that which would normally come from that generated from testing the patient or QC sample; it is used to check the operation of all the processes downstream from the sensor detection process (22).

For many years, manufacturers of qualitative, single-use tests such as those for pregnancy (HCG) and more recently, cardiac markers, have included a built-in control system that indicates whether the testing system has worked correctly. More recently such systems have been incorporated into quantitative tests. An example is a single-use test strip for determination of INR. The strip has three channels, one for the patient sample and two for different levels of internal QC; all three channels operate simultaneously when a patient sample is applied to the strip (23). On-board QC systems such as these provide a more comprehensive check of the whole testing system than does electronic QC, and it is likely that more similarly designed devices that do not require the user to apply additional QC control samples to test strips will appear in the future.

Impact of Technology on Quality Control Procedures

The introduction of some of the features described above has not been without controversy. This is particularly so for single-use devices that obviously cannot be subjected to exactly the same batch QC procedures as those used for multiple-use devices. Thus a subject of continuing debate is how often QC should be performed on single-use devices. Over the years, electronic QC has become more sophisticated. In 1996 CLIA accepted electronic QC, for a particular device and under certain conditions, as a substitute for traditional QC processes involving control samples. However there is still the view that no electronic QC system is capable of measuring the performance of the whole testing process—and therefore electronic QC cannot provide the same information as the measurement of a patient or QC sample provides.

To address these issues, the CLSI established a committee to address quality management for unit-use testing. As a result of its deliberations, CLSI issued a guideline in 2002 (24). Using a systems approach, the document identifies possible errors that can happen in the different testing phases. These errors are documented in a "Sources of Error Matrix" checklist, which can be used by manufacturers, users, and regulatory bodies. Thus manufacturers can use the matrix to identify potential failure modes and lessen the risk of their occurrence, while users can customize their quality management programs according to the errors in the matrix.

The guideline also emphasizes that there is a range of quality management procedures that contribute to the quality of single-use devices, and it is important to consider liquid QC as just one, albeit important, contribution. Some of these quality management processes are shown in Table 2-4.

The introduction of QC-related technological innovations has recently led the major regulatory body in the U.S.—the Centers for Medicare and Medicaid Services (CMS)—to describe some new approaches to QC for POCT devices as part of a general guide for implementing CLIA 88 for all laboratory testing (25). These so-called Equivalent QC Options were designed to reduce the amount of QC testing for those devices that have internal monitoring facilities, including electronic QC, and can demonstrate a degree of system stability. As it happens they have been roundly criticized by all parties, and it seems they are unlikely to be introduced in their present form (26).

The main conclusion to draw from the above discussion is that at the present time there is no consistent agreement on the correct approach to the quality management of single-use devices. No doubt some devices now incorporate quite sophisticated QC procedures, but how these devices operate is not always clearly stated, and users should evaluate claims carefully.

TABLE 2-4. Quality management processes as applied to single-use POCT devices

- Standard operating procedures. Each unit-use test should have a written procedure that covers all aspects of the testing cycle. The procedure should include sources of error that are detected by the operator, dependent on proper technique, or managed by training.
- Training and competency. Traditional liquid QC should be used as a measure of operator competency. Frequent operators (those performing tests at least once per week) would perform liquid (i.e., not electronic) QC at least once per week. Infrequent operators (those performing tests less than once per week) would perform liquid QC with each day of testing. Each institution could modify these recommendations based on data and experience. Reagent stability determines the absolute minimum frequency: quality control testing intervals should be no longer than 1/10th of the stated reagent stability. In practical terms, QC testing should be performed no less than every 4–6 weeks.
- Ongoing process control. The goal of process control is to verify that system components (i.e., operator, instrument, reagents, sample, and environment) are performing as specified by the manufacturer and at a quality level acceptable to the user. Various forms of controls are available to test different parts of the process: acceptance testing; periodic quality control (traditional liquid quality control); split samples; other forms of quality control (e.g., electronic QC); preventive maintenance; proficiency testing; delta checks; environmental monitoring; and clinical surveillance.
- Error and incident reporting. Retrospective review of variations, errors, and problems reported in the testing cycle may be used to improve product design and prevent errors (17).
- Auditing. Prospective searches for problems in the testing cycles can lead to quality improvement or corrective action.

As with previous generations of POCT devices, users should be guided by the reproducibility and overall analytical performance of the system and the number and competence of the operators, together with the frequency with which the system is used. Thus some may wish to continue with a QC testing strategy that is similar to that used for multi-use devices, namely analyze a minimum of one quality control sample per run during each shift. If testing is infrequent, then an alternate approach is to analyze QC whenever there is a change to the testing system, such as a different batch of testing materials or a different operator (27). It is hoped that one of the outcomes of the present debate in the U.S. on Equivalent QC will be some clearer guidelines in this area.

In summary three major factors need to be controlled and supported by QC procedures to ensure the quality of POCT:

- The manufacture of the device and reagents.
- The storage of reagents.
- The competence of the operator.

External Quality Assurance or Proficiency Testing

External quality assurance (EQA) is mandatory for central laboratories throughout the world. This is also becoming true for POCT in many developed countries (28–30). The aim of all EQA schemes is to provide a comparison between users of the same analytical system. Thus EQA provides an opportunity to not only compare POCT systems but also to compare POCT results with the central laboratory. This can be important when patients are managed in several departments—or when machines break down and samples are taken to other sites for testing.

The design of these schemes is similar: samples go to all participants who then return the results after analysis to the organizers who produce a performance report. Traditionally there may have been a delay of some weeks between the analysis and receipt of the report, but now there are Internet-based EQA schemes that offer a much more rapid turnaround time. Users enter their results via a web browser and when a minimum number of results have been entered, users can obtain an immediate report on their performance. Several POCT device manufacturers offer EQA schemes for their users or contract with a third party to provide a similar service. An example of data from an external quality assurance scheme for POCT is shown in Figure 2-2.

Some hospital networks organize their own EQA surveys within their institution. This becomes feasible and indeed can be more practical when a healthcare provider uses multiple devices, such as glucose meters, throughout the network.

One problem shared by all types of EQA schemes is the nature or matrix of the QA specimen. The most common specimen used by POCT devices is whole blood, which is unstable for many measured substances and therefore cannot easily be distributed as EQA survey material. The use of other artificial materials with matrices quite unlike whole blood creates obvious problems for many analytes, including oxygen and glucose. There is no easy resolution to this problem, and users of EQA programs must be aware of possible matrix effects on the EQA results.

When deteriorating or poor performance is identified in EQA schemes, it is important to document the problem and then provide and document a solution. This information should be integrated into the quality record for the POCT site or location. It may be necessary as part of this exercise to go back and review some of the patients' notes to ensure that incorrect results have not been reported and that inappropriate clinical actions have not been taken. In addition, if the solution highlights a vulnerable feature of the overall process, or of one particular operator, then retraining must be instituted.

Maintenance and Inventory Control

Maintaining instruments and adequate supplies of reagents are tasks familiar to all laboratory staff. The challenges with these tasks in relation to POCT are

standard report

report settings

sample	2006-02		n	2864
analyte	Glu		minimum	139
reporting in	mg/dL		maximum	320
deviation	relative (resolution 1%)		average	214
reference method	median		median	216
reference value	216 mg/dL		SD	22
calibration	plasma (recommended)		CV	10.1%

< –25% **39 Result(s)**

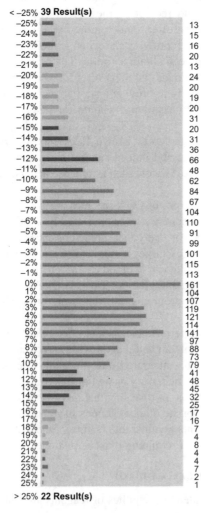

–25%	13
–24%	15
–23%	16
–22%	20
–21%	13
–20%	24
–19%	20
–18%	19
–17%	20
–16%	31
–15%	20
–14%	31
–13%	36
–12%	66
–11%	48
–10%	62
–9%	84
–8%	67
–7%	104
–6%	110
–5%	91
–4%	99
–3%	101
–2%	115
–1%	113
0%	161
1%	104
2%	107
3%	119
4%	121
5%	114
6%	141
7%	97
8%	88
9%	73
10%	79
11%	41
12%	48
13%	45
14%	32
15%	25
16%	17
17%	16
18%	7
19%	4
20%	8
21%	4
22%	4
23%	7
24%	2
25%	1

> 25% **22 Result(s)**

legend:

excellent (<= 10%)

good (> 10% and <= 15%)

fair (> 15% and <= 20%)

unacceptable (> 20%)

FIGURE 2-2. A screenshot from the CueSee interactive website showing data from an external quality assurance program for glucose. The web-based program is used by several major IVD suppliers and EQA organizers to provide external QA services to users of their products.

once again the need to involve non-laboratory staff and the fact that instruments and supplies can be spread over large distances.

Technical maintenance procedures (such as checking the optics) on many single-use POCT devices are relatively simple. If the device fails, it is usually easier to exchange it for another one; the difficulty is often one of recognizing that it is broken or has a fault in the first place. The biggest and sometimes the only maintenance task is removing spilled blood and ensuring that using the instrument does not pose a health and safety hazard. Again, better instrument design is reducing this problem, but even the best designed system is no remedy for poor and sloppy operator technique. The latter must be addressed by rigorous and regular training.

Maintenance of both adequate and correctly stored reagents is a greater challenge. Reagents are increasingly robust but cannot be made to survive climatic extremes or operator carelessness. Thus the use of standard operating procedures that dictate how reagents are to be stored and handled is essential; this is particularly pertinent for reagents that must be refrigerated until they are used for analysis, when they must be equilibrated to room temperature. Manufacturers increasingly find ways to prevent the use of expired reagents, usually through the use of bar code systems, but procedures that continually monitor reagent shelf life must be in place.

Documentation

Like any other laboratory practice, all POCT procedures should be documented. Many patient results produced by POCT are never recorded. This has been one of the key criticisms against this type of testing. Apart from the obvious importance of having a record of patient results, first and foremost in documentation should be a Standard Operating Procedure for the device. Other important records include training and certification of operators; all calibration, quality control, and quality assurance data; and error logs with any corrective actions.

Fulfilling these tasks manually is a major challenge, particularly for busy clinical staff. Fortunately information technology is assisting with these problems. More sophisticated devices have software for automatically recording much of this information, and with the advent of continually cheaper electronic storage, even the smallest devices are likely to include such features. Newer POCT devices, for example, have increasing capacity for storage and memory, so that a wealth of patient and technical data can be accessed after the analysis has been performed. Linking of devices to information systems, so-called "connectivity," will also facilitate the documentation of appropriate information, including getting the patient result into the patient record.

It is also important to document all external quality assurance data, together with operator and instrumentation identification. In this way it is possible to track the performance of systems and in so doing identify failing performance so that trends that might lead to clinically important errors can then be corrected, reducing the risk to patients.

Staff Training and Certification

The confidence of the clinician, the caregiver, and the patient in the results generated by a POCT device depends heavily on the robustness of the instrument and the competence of the operator. Many of the agencies involved in regulating healthcare delivery now require that all personnel associated with the delivery of diagnostic results demonstrate their competence through a process of regulation—and this applies equally to POCT.

As mentioned earlier, healthcare professionals involved in POCT are unlikely to have received training in the use of analytical devices. For nursing staff, this lack of training can be compounded by an attitude that does not regard POCT as a core nursing duty, yet they will be expected to operate what maybe relatively complex pieces of equipment.

Thus an effective training program that motivates as well as instructs personnel to perform POCT is crucial. The key elements of a training program are listed in Table 2-5. Program delivery can include formal group or one-to-one

TABLE 2-5. The key elements of a POCT training program

Understanding the context of the test
- Pathophysiological context
- Clinical requirement for the test
- Action taken on basis of result
- Nature of test and method used

Patient preparation required
- Relevance of diurnal variation
- Relevance of lifestyle, e.g., exercise
- Relevance of drug therapy

Sample requirement and specimen collection

Preparation of analytical device—machine and/or consumables

Performance of test

Performance of quality control

Documentation of test result and quality control result

Reporting of test result to appropriate personnel

Interpretation of result and sources of advice

Health and safety issues
- Specimen collection
- Disposal of sample and test device
- Cleaning of machine and test area

presentations, self-directed learning using agreed-upon documentation, or computer-aided learning (31).

Whatever the training strategy, it is important to document the satisfactory completion of training and the individual's competence. The latter can be assessed at the end of training. This assessment should involve questions that allow individuals to demonstrate both their understanding and their skills. Finally, the operator should be shadowed through the whole procedure involved in the point-of-care test on a minimum of three occasions.

Competence is maintained through regular practice of skills and continuing education, and it is important to build these features into any education and training program. However, the most important thing is to encourage an open approach to the assessment of competence, so that operators can themselves seek help if they feel that problems are occurring. Ideally each POCT location should feature a key trainer or a "champion" who takes ownership of the testing and can deal with at least some of the day-to-day problems. This avoids sending every issue back to POCT Coordinators or their staff.

The open approach should be supported with audit and performance review meetings where problems can be aired and developments discussed. The regular assessment of competence should be built into a formal program for the recertification that is a requirement of most accreditation programs.

Audit

While audit includes the review of analytical performance through both quality control and quality assurance data, it goes beyond QC and QA to review the way POCT is used, and its effectiveness in the wider healthcare setting (32). An example of this is the use of HbA1c testing in the management of diabetes, where the analytical results are reviewed together with the clinical history in the patient files. Though this approach does not target the POCT element directly, it is possible to see if the testing is not being performed, or if the results are not being used properly.

A new form of audit, disease registries, is now being used by patients, caregivers, and healthcare purchasers and providers. These audits show the complete care pathway and document statistics related to the morbidity and mortality of the diseases, as well as sometimes including some surrogate measures such as HbA1c for glycemic control (33, 34).

THE ROLE OF INFORMATION TECHNOLOGY IN POCT

Consider a typical POCT scenario where dozens if not hundreds of blood glucose meters are located in multiple wards of the three or four hospitals in the healthcare institution's network. In addition there are five to ten blood gas analyzers as well as several devices for measuring cardiac markers all located outside and some distance from the central laboratory. The manager responsible for this network must perform all of the many tasks that have been described earlier in order to be confident that a clinically effective POCT service is being delivered. It is not difficult to see that information technology can play a key if not essential role in delivering the service.

Laboratory information systems (LIS) have existed for several decades for the purpose of combining analytical data with other information, such as patient demographics, to produce useful and accessible clinical information. This informatics process goes hand in hand with other LIS processes that are designed to ensure and provide evidence (such as through audit trails and other quality-related statistics) of the quality of all transactions that take place. Such procedures are now an accepted part of laboratory practice, and organizations such as the Joint Commission on Accreditation of Healthcare Organizations in the U.S., and similar regulatory bodies in other countries, demand processes such as real-time quality control.

Providing similar informatics for POCT devices has proved extremely difficult. Some of these information management problems are shown in Table 2-6; the lack of a database within the device has been a key deficiency. Thus, POCT data have often had to be entered manually into an information system, introducing major risks such as transcription error, or more likely, the data never reaching the patient record. Thus important clinical information may be lost, with the additional possibility of costly duplicate testing (35). Newer devices have addressed this problem by incorporating a database, but linking this database to information management systems remains a problem because each device has its own proprietary interface and the costs of interfacing every type of device is prohibitive.

This lack of "connectivity" was regarded as a major barrier to the effective deployment of POCT (36). Accordingly the Connectivity Industry Consortium (CIC) developed a set of seamless "plug and play" point-of-care communication standards that are now maintained as POCT1-A by the

TABLE 2-6. Principles and problems with POCT and information management

Principle	Problem with POCT
Patient samples should be positively identified to the system	Strip devices do not carry identification and meters do not read IDs
Operator access to the system should be password protected	Not true of many hand-held meters
A hard copy record of the results, positively identified, should be generated	Many meters do not carry printers and results must be manually transcribed
Data should be stored in a local or remote database	Not available on the majority of devices
A log of events and of access should be maintained	Not available on the majority of devices

Clinical Laboratory Standards Institute (37). Essentially, if a POCT device incorporates this standard it should easily communicate with laboratory and hospital information management systems, allowing exchange of data and information in a standardized format irrespective of vendor, location, or interface. The existence of the standard introduced the hope that in the next few years electronic connectivity of POCT devices would become commonplace.

The reality has been somewhat different. Few devices as yet incorporate the POCT1-A standard. A recent U.S. survey has shown that there are still large numbers of POCT devices where the results are manually input into the LIS or patient records. However, this situation is likely to improve as legacy devices are replaced by new ones that incorporate the standard. In addition the Clinical and Laboratory Standards Institute (CLSI) is working on guideline documents for both users and manufacturers that will better explain what the standard can do and the benefits it can deliver. This should encourage the purchasers of POCT equipment to demand that the POCT1-A standard be incorporated in the device software.

MAKING THE BUSINESS CASE FOR POCT

The case for the introduction of any new test or testing device is based on the identification of need, the demonstration of clinical utility with evidence from properly conducted outcomes studies, and formal assessments of risk, health, and economic outcomes.

The case for a POCT service can be based on responses to a series of questions that identify the test itself but that should also explain why the current service is not meeting the needs of the patient or the clinician. Specifically the case should identify what benefits are likely to accrue to the patient and/or the healthcare provider by introducing a more rapid service. The case will contain an economic assessment of the costs of delivering the current service and the costs associated with the POCT service. The benefits need to be identified and accounted for in a similar way—although this is more complex as it involves identifying the cost of other elements of the health service, which may not always be amenable to a simple accounting approach.

However it is important in such an exercise to ensure that the costs associated with the benefits are recognized in terms of those that will free up resources, as against those that may be more associated with an improved quality of care and ultimately improved quality of life. Some of the costs in the latter category cannot be readily leveraged, and therefore resources are required to fund the POCT service. This part of the exercise is extremely useful as it often identifies the changes in practice required to release resources for funding the new service; it also identifies the potential difficulties associated with achieving such a change. The risk assessment will focus primarily on the procedures and processes that need to be put in place in order to maintain a high quality of service.

SUMMARY AND RECOMMENDATIONS

- All POCT should be managed by an institution-wide committee with representation of all parties who contribute to or use the service.
- The POCT Committee should utilize the skills and experience of laboratory-trained professionals.
- POCT should be part of the institution's quality management program and include the training and certification of all POCT operators.
- The clinical effectiveness of the POCT service should be reviewed regularly.
- All applications for a POCT service should be supported by a proper business case.

REFERENCES

1. Burnett D. Accreditation and point-of-care testing. Ann Clin Biochem 2000; 37:241–3.
2. Burnett D. A practical guide to accreditation in laboratory medicine. London: ACB Venture Publications, 2002:314pp.
3. Jacobs E, Hinson KA, Tolnai J, Simson E. Implementation, management, and continuous quality improvement of point-of-care testing in an academic healthcare setting. Clin Chim Acta 2001;307:49–59.
4. Pearson M. Equipment procurement and implementation. In: Price CP, St John A, Hicks JM, eds. Point-of-care testing, 2nd ed. Washington, DC: AACC Press, 2004:127–35.
5. U.S. Department of Health and Human Services. Medicare, Medicaid, and CLIA programs: regulations implementing the Clinical Laboratory Improvement Amendments of 1988 (CLIA). Final rule. Federal Register 1992; 57:7002–186.
6. U.S. Department of Health and Human Services. Medicare, Medicaid, and CLIA programs: regulations implementing the Clinical Laboratory Improvement Amendments of 1988 (CLIA) and Clinical Laboratory Act program fee collection. Federal Register 1993;58:5215–37.
7. Ehrmeyer SS, Laessig RH. Regulation, accreditation, and education for point-of-care testing. In: Kost G, ed. Principles and practice of point-of-care testing. Philadelphia: Lippincott Williams and Wilkins, 2002:434–43.
8. International Organization for Standardization. International Standard ISO 22870. Point-of-care testing (POCT) requirements for quality and competence. Geneva: ISO, 2006.
9. Freedman DB. Clinical governance—bridging management and clinical approaches to quality in the UK. Clin Chim Acta 2002;319:133–41.
10. Price CP. Point-of-care testing in diabetes. Clin Chem Lab Med 2003;41:1213–9.
11. St John A, Davis TME, Goodall I, et al. Nurse-based evaluation of point-of-care assays for glycated haemoglobin. Clin Chim Acta 2006;365:257–63.
12. Petersen PH, Fraser CG, Kallner A, Kenney D, eds. Strategies to set global analytical quality specifications in laboratory medicine. Scand J Clin Lab Invest 1999;59:475–585.
13. Skeie S, Thue G, Nerhus K, Sandberg S. Instruments for self-monitoring of blood glucose: comparisons of testing quality achieved by patients and a technician. Clin Chem 2002;48:994–1003.

14. Clinical and Laboratory Standards Institute. Evaluation of precision performance of clinical chemistry devices: approved guideline. CLSI Document EP5-A. Wayne, PA: CLSI, 1999.
15. Bullock DG. Quality control and quality assurance. In: Price CP, St John A, Hicks JM, eds. Point-of-care testing, 2nd ed. Washington, DC: AACC Press, 2004: 137–45.
16. Laessig RH, Ehrmeyer SS. Quality management and administration of point-of-care testing programs. In: Kost GJ, ed. Principles and practice of point-of-care testing. Philadelphia: Lippincott Williams & Wilkins, 2002:422–33.
17. St John A. Benchtop instrument for point-of-care testing. In: Price CP, St John A, Hicks JM, eds. Point-of-care testing, 2nd ed. Washington DC: AACC Press, 2004:31–45.
18. Hirst D, St John A. Keeping the spotlight on quality from a distance. Accred Qual Assur 2000;5:9–13.
19. Westgard JO. What's wrong with traditional quality control? Clinical Biochemist Newsletter 1999;June:13–4.
20. Westgard JO, Fallon KD, Mansouri S. Validation of iQM active process control technology. Point of Care 2003;2:1–7.
21. Nichols JH, Dyer KL, Humbertson SK, et al. Traditional liquid quality control versus internal quality control. Point of Care 2002;1:9–19.
22. Westgard JO. Electronic quality control, the total testing process, and the total quality system. Clin Chim Acta 2001;307:45–8.
23. Jina A. A novel point-of-care prothrombin time monitoring system. Chest 2000;118:Suppl 2835.
24. Clinical and Laboratory Standards Institute. Quality management for unit-use testing: approved guideline. CLSI Document EP18-A. Wayne, PA: CLSI, 2002.
25. Clinical Laboratory Improvement Amendments (CLIA) Equivalent Quality Control Procedures. Brochure #4. Available at http://wwow.cms.hhs.gov/clia/ (Accessed October 20, 2004).
26. Habig RL, Boone J, Yost J, et al. Quality control for the future. Lab Medicine 2005;36:609–40.
27. College of American Pathologists. Available at http://www.cap.org/lap/fudt.html (e.g., drugs of abuse testing program) (Accessed March 10, 2006).
28. United Kingdom National External Quality Assessment Scheme. Available at http://www.ukneqas. org.uk (e.g., cholesterol testing program) (Accessed March 10, 2006).
29. Wales External Quality Assessment Scheme. Available at http://www.weqas.co.uk (e.g., urinalysis program) (Accessed March 10, 2006).
30. Clinical and Laboratory Standards Institute. Point-of-care in vitro diagnostic (IVD) testing: approved guideline. CLSI Document AST2-A. Wayne, PA: CLSI, 1999.
31. Storto Poe S, Case-Cromer DL. Nursing strategies for point-of-care testing. In: Kost G, ed. Principles and practice in point-of-care testing. Philadelphia: Lippincott Williams and Wilkins, 2002:214–35.
32. Barth JH. The role of clinical audit. In: Price CP, Christenson RH, eds. Evidence-based laboratory medicine: from principles to outcomes. Washington, DC: AACC Press, 2003:209–24.
33. Finnish Diabetes Association Development Program for the Prevention and Care of Diabetes in Finland 2000–2010. Available at http://www.diabetes.fi/ english/programme/programme/chapter14.htm (Accessed March 10, 2006).

34. The Renal Association. The UK Renal Registry. Available at http://www.renalreg.
    com (Accessed March 10, 2006).
35. Jones R. Informatics and point-of-care testing. In: Price CP, Hicks JM, eds. Point-
    of-care testing. Washington, DC: AACC Press, 1999:175–95.
36. Stephens EJ. Developing open standards for point-of-care connectivity. IVD
    Technology 1999;10:22–5.
37. Clinical and Laboratory Standards Institute. Point-of-care connectivity: approved
    standard. CLSI Document POCT1-A. Wayne, PA: CLSI, 2001.

Chapter 3

Laboratory Medicine, Point-of-Care Testing, and the Process of Care

Healthcare consists of many different care pathways. These pathways reflect the range and severity of symptoms with which patients present, the diagnosis that is made, the treatment prescribed, and the outcomes achieved. The progress of the patient through the healthcare system can be described as the *patient journey;* a *patient care pathway* is often described as one part of that journey, which in managerial terms is sometimes called *the episode of care* (1). One representation of this journey, the process of care in the care pathway, is represented in Figure 3-1. We use the term *process of care* regularly throughout this book as POCT is about doing things differently, and therefore has an impact on the process of care. An important element of this journey, which may include more than one episode of care, is the use of various tool sets, e.g., diagnostic tests, to assist the caregiver in making decisions, initially on the diagnosis, and subsequently on the choice of therapy and its efficacy, as described in Chapter 1.

A completed episode of care may involve several iterations of these tool sets. Thus in the case of a patient with a long-term (chronic) condition such as diabetes mellitus, the monitoring component will involve constant iterations of

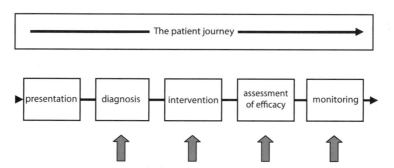

FIGURE 3-1. A high-level view of the care process or care pathway, indicating the points ⬆ at which a diagnostic service may contribute to clinical decision making.

the "diagnosis and intervention" elements as a diagnostic test result is used to assess the efficacy of treatment and to change management accordingly. POCT enables rapid test results, available at the point of care. However, just because rapid test results are possible does not mean that tests need to be conducted at the point of care; it is possible in many situations to meet the need for rapid test results by changing other aspects of the process. For example, the patient could be asked to visit the laboratory to provide a blood sample an hour prior to attending a routine anticoagulation clinic visit, or a technician could transport laboratory equipment to the clinic to make the HbA1c measurements during the diabetes clinic session.

Regardless of which approach is used, it is important to recognize that the appropriate and successful utilization of POCT will often involve a change in the process of care.

THE ORGANIZATION AND MANAGEMENT OF HEALTHCARE

Reviewing some of the basic organizational structures and processes that lie behind healthcare delivery is helpful in understanding the changes that may be required to implement POCT, and the benefits to be derived from it. The traditional view of healthcare delivery involves either clinical settings, or funding mechanisms, as healthcare is generally viewed as a societal cost rather than as a societal responsibility. Figure 3-2 illustrates one common view of the organization of healthcare. A history of the development of any healthcare system will describe the evolution of healthcare and the responsibilities of delivery organizations such as the National Health Service (NHS) in the United Kingdom (2).

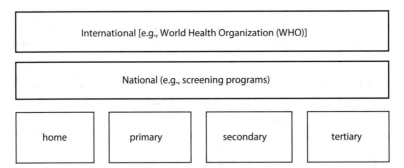

FIGURE 3-2. The various organizational segments of healthcare delivery. Typically the patient journey begins either in the home or primary care setting, but the patient might also enter at the secondary care level with an acute presentation. A crucial issue in any complex service such as healthcare is communication, especially when healthcare delivery is made through a number of organizations. In a country providing so-called "socialized medicine" there will be some line of accountability among all services provided at a national level.

Societal Aspects of Healthcare

In many countries healthcare is provided with some degree of intergovernmental collaboration, especially in relation to the spread of infectious diseases and prevention of environmentally related diseases. At the international level, the World Health Organization plays an overarching role: its objective is "the attainment by all peoples of the highest possible level of health" (3). At a national level both funding of care and societal commitment are organized in a variety of ways. Thus in most countries certain agencies perform some form of public health function, covering surveillance for outbreaks of disease (particularly infectious diseases) as well as lifestyle guidance. Some countries, such as the UK, provide national screening programs for neonatal inborn errors such as phenylketonuria and hypothyroidism, for conditions associated with pregnancy such as Down syndrome, and for early detection of cancers of the cervix, colon, and breast.

Healthcare is usually funded through general taxation, in the case of so called "socialized medicine" such as the National Health Service (NHS) in the UK; or via government regulation through insurance; or through individual payment. If no national program exists, there may be coverage for the sector of the population that is unable to afford insurance, e.g., Medicare and Medicaid in the United States. The allocation of resources for individual aspects of healthcare (e.g., home, primary, and secondary care) varies, as does the interrelationship between these care sectors. Thus in the case of laboratory medicine, funding is generally either through a block allocation from a healthcare purchaser based on the previous year's budget or through a system of reimbursement based on some nationally agreed reimbursement schedule. Quality standards for healthcare and diagnostic services may also be established at a national level.

Healthcare Settings

Home care

Some care takes place in the home (and workplace), for example, including the management of long-term diseases such as diabetes mellitus. The home or workplace is also the setting where health education is considered to be most effective because of the opportunities to influence lifestyle (diet and exercise). It is also the setting targeted by numerous Internet sites dedicated to healthcare support, including Lab Tests Online (4) and NHS Direct (5).

Primary care

Until recently, most peoples' first point of contact with any form of healthcare provision was primary care (with the exception of an emergency). The main contact is with a primary care physician, but other contact points are developing, including telephone hot lines such as NHS Direct, and community pharmacists. The advent of the Internet has enabled individuals to make more use of this facility to access healthcare information.

While the spectrum of clinical presentations varies quite significantly depending on the type of people represented in the population, there are few formal data on the presenting symptoms that prompt a first visit to the primary care physician. Though a number of presentations are due to self-limiting conditions such as sore throats, a high proportion of presentations is associated with long-term conditions, and appropriately there is a trend to manage conditions such as diabetes, asthma, and heart failure in the primary care setting (6). Thus there will be a limited amount of testing performed in this setting for disease diagnosis, but more will be undertaken as part of long-term disease management.

The community pharmacy plays an important role at the primary care level, and it may be the first port of call in the case of minor conditions. However, long-term management of health conditions is also taking place in the community pharmacy setting as the pharmacist becomes more closely integrated into the healthcare delivery team (7).

In some countries the community hospital is organized and managed as part of the primary care sector. It, too, may take responsibility for long-term care, especially for the elderly, and perform more acute care in rural settings.

Secondary care

The acute hospital with an emergency department represents the typical secondary care setting. The work of the hospital typically includes acute and emergency care as well as a mixture of elective and non-elective procedures. Secondary care also includes ambulatory care (outpatient clinics), which in most cases constitutes the highest proportion of episodes—typically about 75% of the total annual attendances. The ambulatory care workload includes a high proportion of long-term disease management, and where this is shared with primary care, it will cover the more complex and advanced cases.

As a result of the acute nature of some clinical work, all secondary hospitals will include diagnostic facilities, and in many countries this can include a laboratory serving both the primary and secondary care sectors.

Tertiary care

This is a hospital setting typically found in the more highly populated urban area, where it serves a larger population. Tertiary care units operate as specialist units within a secondary care setting, or as stand-alone units. Examples include specialist hospitals for children, or for adults with cancer, cardiology, or infectious diseases.

Tertiary care units are supported by diagnostic units, and invariably also have significant research facilities. Primary, secondary, and tertiary care facilities are often linked with universities that incorporate undergraduate and postgraduate education of healthcare professionals.

ORGANIZATION OF SECONDARY CARE AND LABORATORY MEDICINE SERVICES

The first diagnostic laboratories were located in hospitals and were managed within the overall organizational framework of the hospital, a situation that still persists today in many countries of the world. The traditional organization chart describes the way that most hospitals (secondary care sector) are managed (Figure 3-3). The departments are functional groups, and there may be a number of these within the departmental groupings. The laboratory medicine group, for example, may include blood sciences (clinical biochemistry, hematology, and immunology); cellular pathology (anatomical pathology [histopathology] and cytopathology); and infectious diseases (microbiology and virology) (8,9). In the case of medicine it may contain, for example, departments of general medicine, diabetes, endocrinology, cardiology, oncology, and renal medicine. The number of departments determines the size of the management board of the organization, and its day-to-day operations are undertaken by an executive group typically made up of the chief executive, the medical director, the nursing director, the finance director, the human resources director, and an operations director. Thus the management board might include this executive group, plus the directors of each of the medical specialty departments.

Generally speaking, managers who see their organizations as a vertical framework manage in a vertical manner. This equally applies to most healthcare organizations. Thus managers will develop their own departmental standards and working protocols in order to best deliver the objectives for which

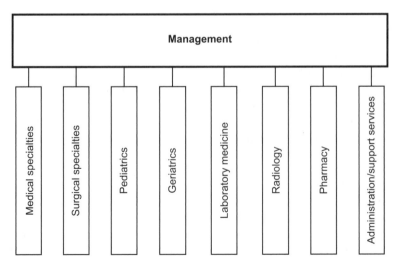

FIGURE 3-3. An example of a traditional organizational chart for a secondary care provider. The figure indicates the relationship between management and the various delivery and support functions of the organization, including diagnostic services. If the latter were provided by an independent organization, the management relationship would be similar in relation to its accountability to the primary organization.

they are responsible. This approach, often called "silo thinking" and "silo management," leads managers to make their departments as efficient as possible, without necessarily taking into account what is happening in the other departmental silos (10), or in the organization as a whole in both the general and specific aspects of healthcare delivery (11–15).

In silo management, managers identify who reports to whom and set goals or targets for each group independent of the others. When cross-departmental (often regarded as cross-boundary) issues arise, they are referred up the management chain until they reach a manager who is responsible for the work performed in both of the departments involved. This management style has been one of the key issues in laboratory medicine over the years, because of its impact on quality, efficiency, and benchmarking (16,17).

This vertical framework also determines the financial flows (and targets) within the organization, and consequently leads to "silo budgeting" and an undue focus on the cost of care rather than on the value to be gained, a point recognized in recent reports from the Health Select Committee, Department of Trade and Industry, and Department of Health in the UK. These reports concluded that there is little incentive for developing and implementing new technology in healthcare (18–20). A similar culture exists with respect to POCT—there is an undue emphasis on the cost of the test, rather than on the value of the benefit (21).

Healthcare has become increasingly "managed," partly because of a desire to control quality and costs, and also to ensure equality of healthcare delivery especially in countries with socialized medicine systems (22). As a consequence, the management structure has become increasingly complex. Patients who present initially to primary care may be seen by several physicians and surgeons in order to obtain a diagnosis, or to obtain treatment in the case of disease progression. Consider the patient with diabetes, who might be involved with both primary and secondary care sectors within the same organization, or even in different organizations. Can management meet the challenge of providing sufficient resources, not only for the correct level of care provided within the organization's primary and secondary sectors, but also for the increasing number of patients within the population for which they are responsible in terms of the overall delivery of care? In the vertical silo, the management must apportion resources correctly between different processes and departments (e.g., staffing in hospital beds and clinic sessions, tests from the laboratory, and drugs from the pharmacy). Management must also manage budget and workload targets for their individual departments as well as for the organization as a whole.

Diagnostic services, such as laboratory medicine, are invariably managed as an independent "silo" or as part of a "diagnostics silo." The laboratory's resource allocation is based on workload and some form of reimbursement; the resource allocation may be on a block contract from the purchaser—a local health authority in the case of a socialized medicine environment—or it may be related to workload. The laboratory services are managed in a similar manner based on targets related to work load and efficiency. In an era of clinical governance the targets include the quality of the result and subsequent interpretation, as well as measures such as turnaround time. There is little attempt to relate the use of the laboratory services to the health outcomes of the patients for whom the tests are being provided.

In some countries the laboratory medicine service, or part of it, is outsourced to an independent supplier; this may also be true of clinical services. Typically investigations requiring a rapid response (those with a turnaround time requirement of less than four hours) will be performed on site in the hospital, whereas those investigations that are not required for immediate clinical management are sent to an offsite facility. The off-site workload is usually of three types: (1) highly specialized investigations (and referrals for a second opinion); (2) workload for which there is not a high demand (i.e., local investment in that investigation is not justified); and (3) work from primary care. Laboratory services purchased from an independent supplier are managed in exactly the same way as the "in house" service, namely as part of a diagnostics silo.

THE PROCESS OF OBTAINING DIAGNOSTIC INFORMATION

The process of obtaining diagnostic test information is summarized in Figure 3-4, and broadly comprises preanalytical, analytical, and post-analytical phases.

Preanalytical Phase

The preanalytical phase includes all activities that take place from the time a test is requested to the time the sample is placed in a device for analysis.

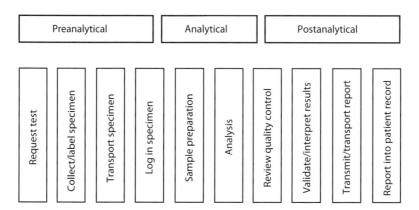

FIGURE 3-4.   A description of the steps involved in generating diagnostic information in the central laboratory.

The activities involved in this phase can include (in the case of a blood sample) the following:

- Identifying the correct test required.
- Determining whether specific collection conditions are required, e.g., fasting (such as for cholesterol measurement); time of day; time after last dose of drug (such as for digoxin measurement); resting; and posture.
- Writing the request form, and including relevant clinical information.
- Delivering the request form to the phlebotomist.
- Identifying the correct collection tubes for blood, etc.
- Collecting specimens.
- Labeling the specimen tubes.
- Sending the specimen and request form to the collection point for delivery to the laboratory.
- Transporting the specimen to the laboratory.
- Logging in the specimen and request information.

Analytical Phase

The analytical phase involves the preparation of the sample, the analysis itself, and the quality measures associated with ensuring that the analysis is reliable and the result can be assured.

One of the major bottlenecks in the laboratory is the removal of the cellular component of blood prior to analysis of serum or plasma. The analytical process may involve more than one instrument or work station, thereby requiring the sample to be separated into a number of aliquots or moved from station to station following analysis. The analytical processes may be automated and perform a number of assays (for example, a panel of liver function tests), whereas others may be more manual. The assay times may vary from a few minutes to several hours—or even days in the case of a microbiological culture. On completion of the analysis, the quality control of the batch of analyses is checked before release of the results for reporting. The phase is completed with the recording of the result on a worksheet, either in paper or electronic form.

Post-Analytical Phase

The post-analytical phase begins with any further validation of the results together with any interpretation that is required. This phase includes the following:

- Comparing the result against reference ranges for each analyte in relation to the population that describes the patient (e.g., child, female, Caucasian, etc.).
- Immediately telephoning the results to the requesting health professional, for example in the case of grossly abnormal results, which may indicate a life-threatening situation and for which there will be defined action ("panic") values.

- Interpreting the results (in the case of abnormal results, or in situations where there has been a significant change in results since the previous analysis). Typically a primary care physician will welcome this additional support; a specialist physician may not require this service. Such interpretive comments will take into account the clinical information that has been provided by the requestor, and may offer guidance on further investigations and/or treatment options.
- Providing interpretive comments (in the case of a large proportion of the microbiological and virological investigations, and in the case of all cellular and molecular pathology investigations).

The diagnostic information, i.e., results plus comments, is then transmitted to the requestor either electronically or in hard (paper) copy. The data are stored on the laboratory information system, which may also provide cumulative reports. Increasingly the information is being stored within electronic patient records.

If one looks at the process of performing a diagnostic test in the context of a classical organizational chart, then it is clear that a number of functional groups, rather than only the clinician or the health professional making the initial request, are involved in the completion of the task. For example, these "groups" may include the health professional(s) who record the request, the phlebotomy service, the specimen transportation service, the information technology department, and again a transportation service if a paper copy of the report is sent back to the requestor. A different person may manage each of these functions, and consequently there are challenges in linking these groups to achieve an efficient and timely transfer of information.

A SYSTEMS VIEW OF HEALTHCARE:
A PATIENT-CENTERED SERVICE

An alternative to the silo organization described above is to take a systems approach. This approach describes the caregiving process from beginning to end, with the input being a request made at the time of a patient consultation (or in expectation of such an event) and the output being the benefit gained from the use of the test result. Taking the systems approach makes it simpler to identify the opportunity for testing at the point of care, and hence to identify potential organizational benefits. This system also puts the focus on the patient.

Two reports—To *Err is Human* (23) and *Crossing the Quality Chasm* (24), published by the Institute of Medicine in the United States in 2000 and 2001 respectively—reflect this focus on the patient. The first report highlighted the number of failures in the system considered to be due to "a disconnected process"; these failures were estimated to have caused nearly 100,000 deaths per year. The second report drew attention to the gap between technology, knowledge, and investment on the one hand and the quality of care on the other. It went on to develop a vision of healthcare in the future—a vision that included the need for a more patient-centered approach. As a result, several organizations initiated a project that led to the report *Building a Better Delivery System* (25), which in part considers a systems approach to the organization and management of healthcare services. In introducing the application of systems engineering in healthcare, one of the authors points out that systems engineering involves "the design, implementation, and control of interacting components or subsystems"—the goal being for the components to work together, in an integrated fashion, rather than to work alone, in parallel or serial configurations.

Thus a systems approach requires thinking more strategically and holistically. In particular it requires *optimizing the patient's journey* (which includes individual care pathways) rather than optimizing the function of individual departments within the organization. This requires managers to visualize and document the patient's path through the organization and understand how the organization works to facilitate efficient (and effective) progress. Though the patients' needs are primary, it must also be recognized that there are a number of stakeholders whose prime interest is how the goals of the healthcare organization are achieved. This is discussed in more detail in Chapter 4.

In business terms, the systems approach focuses on the inputs and outputs of the system. At an operational level, this is embraced within the concept of a care pathway (26). A care pathway incorporates plans that detail the essential steps in the care of individual patients (the input) and describe the expected progress of the patient (the output or outcome). A care pathway is built on the evidence that is summarized in a clinical guideline, interpreted for the individual patient and taking into account the patient's preferences (27). The care pathway equates to the value chain in business. It is interesting in the context of this approach to note that organizations involved in the reimbursement or procurement of the diagnostics elements of healthcare are now seeking evidence on "outputs" or "outcomes" to support resource allocation. For example, the Centers for Medicare and Medicaid (CMS) created a document that describes factors it considers important when deciding whether to provide national coverage for certain medical items and services. These factors are influenced by the amount of evidence showing that a prescribed treatment or medical device is reasonable and necessary. This kind of linkage between coverage and evidence-based data collection had been missing from previous deliberations about medical device or treatment procurement and reimbursement (28). Linking coverage to data collection about medical efficacy is based on the principles of evidence-based medicine, and the goal is the demonstration of improved health outcomes.

On September 1, 2005, the United Kingdom's NHS Purchasing and Supplies Agency (PASA) announced the setting up of a "Centre for Evidence-Based Purchasing" at the time that the Device Evaluation Service transferred from the Medicines and Healthcare Products Regulatory Agency. This followed a recommendation made by the Healthcare Industries Task Force to forge closer links between product evaluation and purchasing. The intention is to "underpin purchasing decisions with objective evidence to support the uptake of useful, safe, innovative products and related procedures in health and social care" (29).

The systems approach toward healthcare (Figure 3-1), including the role of laboratory medicine, provides a great deal more information about, and therefore a better understanding of, the process of care than the traditional view of organizations given in Figure 3-3. In particular it enables the role of the laboratory medicine service to be understood more clearly in terms of what decisions can be made with the aid of the information the laboratory provides. Furthermore, a systems view focuses on process and connections, and therefore ultimately on adaptation and modification, such as might be achieved from the introduction of a new technology in the form of testing at the point of care.

ALTERNATIVE APPROACHES TO RAPID DELIVERY OF RESULTS

Any investigation of the feasibility of POCT should also consider other ways to achieve rapid delivery of results. Where there is an acute clinical need, getting diagnostic information rapidly from a central laboratory requires (1) a rapid transport or vacuum tube system for the specimen and/or a laboratory that is located within easy access of the requestor, together with (2) the availability of electronic reporting or a telephone. However, it is worth noting that in one audit of electronic reporting of urgent results, a high proportion of the results was never accessed (30), suggesting that simply making results available at a reporting terminal is not always sufficient to ensure that the information gets attention.

Collinson et al. (31), in looking at the use of a rapid service for the delivery of cardiac marker results to the coronary care unit, suggested that POCT at the bedside enabled the clinician to complete the consultation by taking appropriate action on the test result instead of having to interrupt the consultation with another patient consultation while awaiting the test results, thereby minimizing the risk of not accessing and acting on the results. If a laboratory cannot provide a rapid turnaround for results when they are urgently needed, then the only alternative to POCT is to locate a testing station close to the clinical settings where the service is required. The testing station implies that it is manned by laboratory-trained staff, whereas POCT implies using other members of the clinical team to perform the tests.

If the immediacy of diagnostic information is justified on the basis that it enriches the patient-clinician consultation, there are a number of options for achieving this immediacy. One of the simplest is to ask the patient to attend a phlebotomy/sample collection center a few days before the consultation for the specimen to be collected/received. Though this is convenient for the laboratory, it cannot be regarded as a patient-centered approach, since it is certainly not convenient for the patient, who, for example, may also require time away from work. Alternatively, if the analyte is relatively stable (e.g., HbA1c), one can send the patient a sample collection device and ask the patient to make the collection and send the device back to the laboratory (32). The third option is to take the testing system to the clinic (as suggested above) and arrange the clinic schedule so that the analyses can be performed before the consultation takes place (33). Each of these alternatives has its limitations either in terms of the inconvenience to the patient, the applicability and reliability of the approach, or the efficiency and productivity of the alternative.

INTEGRATING POCT INTO THE CARE PATHWAY

Having excluded the alternatives described above and decided that POCT is the best option for the service in question, one then must consider how best to integrate it into the care pathway. Three examples will be described to demonstrate the impact of introducing a POCT service. The challenge of managing healthcare is in understanding the elements of these processes, their grouping into manageable units, and the process changes involved when introducing POCT. The central goal is to maintain and improve the "connectivity" of the care pathway and in so doing achieving the desired outputs from the process, namely improved health outcomes.

The outputs should benefit the patient, the clinician or healthcare professional caring for the patient, and/or the healthcare provider organization. If the outputs meet these criteria, they will obviously benefit the healthcare purchaser. With these as the stated objectives, an analysis of the systems approach to healthcare provision through the review of individual patient care pathways will quickly demonstrate the potential utility of POCT.

Example 1: Non-Acute First Presentation to Primary Care

This scenario could apply to any first, relatively non-acute presentation to a primary care physician in which a diagnostic test can help "rule in" or "rule out" an immediate diagnosis. Thus in a case of suspected urinary tract infection, the patient presents describing one or more of a number of symptoms, which in themselves enable the physician to establish the likelihood (the pre-test probability) of a urinary tract infection as the diagnostic hypothesis. A simple urinalysis enables the post-test probability to be established (by whether the test is positive or negative) and thus a decision to treat (or not) with antibiotics to be made (34,35). If the test is performed in a laboratory, then the patient is likely to need two visits to the primary care physician; using POCT would enable this to be accomplished in a single visit. The logistics of the scenario are illustrated in Figure 3-5a, with the upper section showing the laboratory approach and the lower, the POCT option.

The same rationale could apply in the case of suspected Strep A or influenza infections and to screening for *Chlamydia* infection (POCT devices are available for all of these diagnoses). In the latter case the patient typically attends the clinic to provide a sample followed by a second visit to receive the result and the prescription for treatment in the event of a positive result. Audit of this scenario has shown that a large proportion of patients never make the second visit, and then the test is wasted (36).

Example 2: Acute Presentation to the Emergency Department

This scenario could apply to any situation in which the condition may be life threatening and where the test result will establish the diagnosis and guide treatment. For example, suppose a patient presents with chest pain. Biochemical cardiac markers, together with an electrocardiogram (ECG), will be used to make or refute the diagnosis, as well as to assess the prognosis. A positive diagnosis will indicate that an intervention commence, including the possibility of the patient being admitted to a coronary care unit. At a later stage when the acute event has abated, the patient's risk of a further cardiac event will be assessed (e.g., with the aid of cholesterol measurements and assessment of

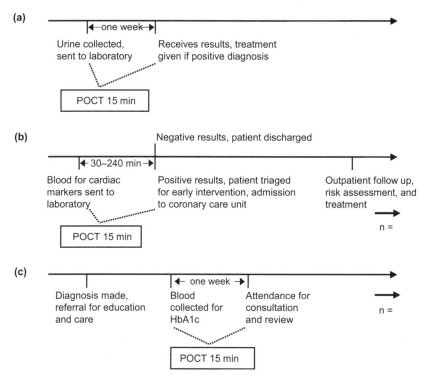

FIGURE 3-5. Representation of three patient journeys: (a) suspected urinary tract infection presenting to primary care; (b) suspected myocardial infarction presenting to the emergency department; and (c) diagnosis of diabetes mellitus, beginning long-term disease management. The down bar represents an episode of care, for example, a visit to a clinic; "n =" denotes that the visit will be repeated. In each scenario the transition from a laboratory to a POCT modality of testing is shown, illustrating the reduction in the length of the process.

coagulation status). The results of this assessment guide long-term drug therapy and the need for lifestyle changes. Following discharge from hospital, the patient is then likely to be monitored at regular intervals, perhaps first in an ambulatory secondary care unit and then in primary care. This process is summarized in Figure 3-5b, with the laboratory-based service shown in the upper portion, and the POCT in the lower.

For the first stage of this pathway, POCT's rapid provision of results at the time of admission assists rapid triage of the patient, reducing length of stay in the emergency department. In the absence of POCT, the specimen must be transported to the laboratory, and following analysis, the results transmitted back to the emergency department. Lee-Lewandrowski et al. (37; see also Chapter 5) showed that using a POCT service reduced the turnaround time for cardiac markers from 110 min to 17 min. Two other studies, Ng et al. (38) and McCord et al. (39), were able to demonstrate a successful "rule out" of myocardial infarction within 90 minutes of admission, which also led to a significant reduction in the number of patients referred to the coronary care unit. The logistics of the scenario are illustrated in Figure 3-5b.

A number of other examples fit the above scenario in the emergency department, including suspected acute poisoning with acetaminophen (paracetamol) or with alcohol, suspected deep vein thrombosis, and suspected hypo- or hyperglycemia. Examples in other acute care units such as the intensive care unit and the operating room setting also exist (40).

Example 3: Long-Term Condition (Disease) Management

This scenario could apply to any situation in which diagnostic tests might be used to manage the efficacy and compliance with treatment for a long-term condition, as in the case of a patient with diabetes mellitus. An initial urine test for glucose heightens suspicion of the condition and prompts a blood glucose test to establish a diagnosis. A positive diagnosis indicates the need to institute therapy relative to the level of glycemic control. This means insulin for a patient with type 1 diabetes or a type 2 patient with very poor control, and diet and/or oral hypoglycemic drugs for type 2 patients with better glycemic control. A protocol for managing the disease is then established, the regime depending on the quality of glycemic control. The protocol includes (1) self-testing for blood glucose to monitor the efficacy of therapy and guide any short-term changes; (2) use of HbA1c measurements to give a longer-term assessment of glycemic control (over the past three months); and (3) monitoring for indicators of risk associated with the complications of diabetes, e.g., cholesterol for cardiovascular disease and urine albumin for renal disease. The latter two steps of the protocol are undertaken initially in an ambulatory care clinic but later devolve to primary care. This process is summarized in Figure 3-5c with the use of the laboratory-provided test results shown in the upper portion and POCT in the lower portion.

In the long-term monitoring stage, the use of a laboratory service may require two visits for each review—one to provide the blood sample, and the second for the consultation—unless both can be completed at the same visit. It has been shown that patients who know their own HbA1c level have better outcomes (41); furthermore, those who are informed of their current HbA1c level at the time of the review consultation appear to fare better (42). Thus POCT for HbA1c can lead to better outcomes, and save a clinic visit.

SUMMARY AND RECOMMENDATIONS

- The increasing emphasis on a patient-centered approach to the delivery of healthcare requires a shift from the classical approach of isolating laboratory medicine services from the complete clinical care pathway to one that offers a more systems-orientated approach.
- Moving away from so-called "silo thinking" requires managers and policymakers to understand the various steps in the patient care pathway, to optimize the process of care in clinical, organizational, and economic terms.
- A systems approach to the organization and management of care will enable the manager or policymaker to make better-informed decisions about the potential role and utility of POCT.
- The key purpose of POCT is to deliver results and diagnostic information quickly, so that decisions can be made more quickly, and so that the actions that follow improve health outcomes.
- A number of stakeholders are involved in each care pathway; the views and expectations of all parties should be understood and taken into account.
- It is important therefore to understand the process changes that may be required and the improved outcomes that can be gained.

REFERENCES

1. Muir Gray JA. The resourceful patient. Oxford: eRosetta Press, 2002:154pp.
2. Rivett G. From cradle to grave: fifty years of the NHS. London: King's Fund Publishing, 1997:506pp.
3. World Health Organization. Constitution. Available at http://policy.who.int/cgi-bin/om_isapi.dll?hitsperheading=on&infobase=basicdoc&jump=Constitution&softpage=Document42#JUMPDEST_Constitution (Accessed March 12, 2006).
4. American Association for Clinical Chemistry. Lab Tests Online. Available at http://www.labtestsonline.org (Accessed March 12, 2006).
5. National Health Service. NHS Direct. Available at http://www.nhsdirect.nhs.uk (Accessed March12, 2006).
6. Secretary of State for Health. Our health, our care, our say: a new direction for community services. London: The Stationery Office, 2006:227pp.
7. Wermeille J, Bennie M, Brown I, McKnight J. Pharmaceutical care model for patients with type 2 diabetes: integration of the community pharmacist into the diabetes team pilot study. Pharmacy World Science 2004;26:18–25.
8. Kricka LJ, Parsons D, Coolen RB. Healthcare in the United States and the practice of laboratory medicine. Clin Chim Acta 1997;267:5–32.

9. Price CP, Barnes IC. Laboratory medicine in the United Kingdom: 1948–1998 and beyond. Clin Chim Acta 1999;290:5–36.
10. Harmon P. Business process change: a manager's guide to improving, redesigning, and automating processes. Amsterdam: Morgan Kaufmann Publishers, 2003:528pp.
11. Paine LA, Baker DR, Rosenstein B, Pronovost PJ. The Johns Hopkins Hospital: identifying and addressing risks and safety issues. Jt Comm J Qual Saf 2004; 30:543–50.
12. Marren JP, Feazell GL, Paddock MW. The hospital board at risk and the need to restructure the relationship with the medical staff: bylaws, peer review, and related solutions. Ann Health Law 2003;12:179–234.
13. Schwermann T, Greiner W, v d Schulenburg JM. Using disease management and market reforms to address the adverse economic effects of drug budgets and price and reimbursement regulations in Germany. Value Health 2003;6 Suppl 1:S20–30.
14. Fazzi RA. Silo management: it doesn't work! Caring 1999;18:62–3.
15. Friedman BA. Orchestrating a unified approach to information management. Radiol Manage 1997;19:30–6.
16. Nevalainen D, Berte L, Kraft C, et al. Evaluating laboratory performance on quality indicators with the six sigma scale. Arch Pathol Lab Med 2000;124:516–9.
17. Price CP. Benchmarking in laboratory medicine: are we measuring the right outcomes? Benchmarking 2005;12:449–66.
18. Department of Trade and Industry. UK sector competitiveness analysis of six healthcare equipment segments. Available at http://www.dti.gov.uk/files/file10462.pdf? pubpdfdload=05%2F1014 (Accessed March 10, 2006).
19. House of Commons Health Committee Report. The use of new medical technologies within the NHS. Available at http://www.publications.parliament.uk/pa/cm200405/ cmselect/ cmhealth/398/39802.htm (Accessed March 13, 2006).
20. Health Industries Task Force (HITF) report. Available at http://www.abhi.org.uk/ patients/htif/ default.aspx (Accessed March 12, 2006).
21. Marshall DA, O'Brien BJ. Economic evaluation of diagnostic tests. In: Price CP, Christenson RH, eds. Evidence-based laboratory medicine: from principles to outcomes. Washington, DC: AACC Press, 2003:159–186.
22. Simonet D. Where does the U.S. experience of managed care currently stand? Int J Health Plann Manage 2005;20:137–57.
23. Institute of Medicine. Kohn LT, Corrigan JM, Donaldson MS, eds. To err is human: building a safer health system. Washington, DC: National Academies Press, 2000:287pp.
24. Institute of Medicine. Crossing the quality chasm: a national health system for the 21st century. Washington, DC: National Academies Press, 2001:383pp.
25. National Academy of Engineering and Institute of Medicine. Reid PP, Compton WD, Grossman JH, Fanjiang G, eds. Building a better delivery system. Washington, DC: National Academies Press, 2005:262pp.
26. Campbell H, Hotchkiss R, Bradshaw N, Porteous M. Integrated care pathways. BMJ 1998;316:133–7.
27. Guyatt GH, Haynes RB, Jaeschke RZ, et al. User's guide to the medical literature: XXV. Evidence-based medicine: principles for applying the user's guides to patient care. Evidence-Based Medicine Working Group. JAMA 2000;284:1290–6.
28. Centers for Medicare and Medicaid Services. Coverage with evidence development. Available at http://www.cms.hhs.gov/coverage/download/guidanceced.pdf (Accessed March 12, 2006).

29. NHS Purchasing and Supplies Agency. Centre for Evidence-Based Purchasing. Available at http://www.pasa.nhs.uk/evaluation/whatwedo (Accessed March 12, 2006).
30. Kilpatrick ES, Holding S. Use of computer terminals on wards to access emergency test results: a retrospective audit. BMJ 2001;322:1101–3.
31. Collinson PO, John C, Lynch S, et al. A prospective randomized controlled trial of point-of-care testing on the coronary care unit. Ann Clin Biochem 2004; 41:397–404.
32. Holman RR, Jelfs R, Causier PM, et al. Glycosylated haemoglobin measurement on blood samples taken by patients: an additional aid to assessing diabetic control. Diabet Med 1987;4:71–3.
33. Grieve R, Beech R, Vincent J, et al. Near patient testing in diabetes clinics: appraising the costs and outcomes. Health Technol Ass 1999;3:1–74.
34. Deville WL, Yzermans JC, van Duijn NP, et al. The urine dipstick test useful to rule out infections: a meta-analysis of the accuracy. BMC Urol 2004;4:4.
35. St John A, Boyd JC, Lowes AJ, Price CP. The use of urinary dipstick tests to exclude urinary tract infection: a systematic review of the literature. Am J Clin Path 2006 (accepted for publication).
36. Gift TL, Pate MS, Hook EW, Kassler WJ. The rapid test paradox: when fewer cases detected lead to more cases treated: a decision analysis of tests for Chlamydia trachomatis. Sex Transm Dis 1999;26:232–40.
37. Lee-Lewandrowski E, Corboy D, Lewandrowski K, et al. Implementation of a point-of-care satellite laboratory in the emergency department of an academic medical center. Impact on test turnaround time and patient emergency department length of stay. Arch Pathol Lab Med 2003;127:456–60.
38. Ng SM, Krishnaswamy P, Morissey R, et al. Ninety-minute accelerated critical pathway for chest pain evaluation. Am J Cardiol 2001;88:611–7.
39. McCord J, Nowak RM, McCullough PA, et al. Ninety-minute exclusion of acute mycardial infarction by use of quantitative point-of-care testing of myoglobin and troponin I. Circulation 2001;104:1483–8.
40. Price CP, St John A, Hicks JM, eds. Point-of-care testing, 2nd ed. Washington, DC: AACC Press, 2004:488pp.
41. Levetan CS, Dawn KR, Robbins DC, et al. Impact of computer-generated personalized goals on HbA1c. Diabetes Care 2002;25:2–8.
42. Cagliero E, Levina E, Nathan D. Immediate feedback of HbA1c levels improves glycemic control in type 1 and insulin-treated type 2 diabetic patients. Diabetes Care 1999;22:1785–9.

Chapter 4

Process Change and Outcomes: A Patient-Centered Approach

In the previous chapter we pointed out that a systems approach was necessary to fully understand a complex organization such as healthcare. We also recognized that any organization has a number of stakeholders, and it is important to understand their needs and expectations.

In a patient-centered approach to healthcare, the "care pathways" are processes that contribute to some of the complexity of healthcare delivery. The key stakeholders in any care pathway will include the patient, clinician and caregivers, and other health professionals (e.g., laboratory medicine professionals, the person ultimately responsible for managing the provider organization, and the purchaser).

A key aim of a systems approach is to understand how the stages in a complex process interact with each other, so that like processes can be grouped together, thereby achieving a critical mass in terms of both quality (effectiveness) and economy (efficiency).

FIRST STEPS TO USING A SYSTEMS APPROACH TO PATIENT-CENTERED HEALTHCARE

The first step of any systems analysis in the business sector is to identify inputs, process, and outputs (1). For healthcare the generic inputs can be represented by the needs of the stakeholders. The outputs will also be stakeholder dependent—albeit there will be overlap in both cases. Certain generic outputs are relevant to all stakeholders: morbidity, mortality, time, and money. The interpretation of each of these is specific to each stakeholder (2). Figure 4-1 summarizes examples of inputs and outputs for stakeholders.

In order to simplify this discussion, we will focus on a single patient care pathway, because the objective of this book is to describe opportunities for using POCT. In an individual care pathway, the key input is a patient who presents with a clinical problem; the output is a satisfactory resolution of that clinical problem, usually called an "outcome."

In the case of a diagnostic service like laboratory medicine, the primary input can be expressed as a clinical question that reflects the immediate clinical

Outcome	Patient	Clinician/Caregiver	Provider	Purchaser
Clinical				
Reduced HbA1c	★★★	★★★	★	★
Reduced cholesterol	★★★	★★★	★	★
Reduced complication rate	★★★	★★★	★★	★★
Reduced recurrence rate	★★★	★★★	★★	★★
Improved mobility	★★★	★★	★★	★★
Patient satisfaction	★★★	★★	★★	★★
Operational				
Reduced operation time	★★	★★★	★★	★★
Reduced clinic visits	★★	★★	★★★	★★★
Conversion to day case	★★	★★	★★★	★★★
Reduced length of stay	★★	★★	★★★	★★★
Transfer of care to primary care	★★	★★	★★★	★★★
Rapid ED triage	★	★	★★★	★★★
Economic				
Reduced blood product usage	★	★	★★★	★★★
Reduced drug usage	★	★	★★★	★★★
Reduced cost per episode			★★★	★★★
Reduced bed requirement			★★★	★★★
Reduced staff and building requirement			★★★	★★★

FIGURE 4-1. Some examples of surrogate outcome measures that can be used in assessing the utility of POCT, but that can also be used as outputs by the major stakeholders in healthcare, indicating the potential variations in importance for the different stakeholders.

presentation or problem. Here is an example: "Are this patient's symptoms of chest pain consistent with having suffered a myocardial infarction?" The output from the laboratory is the provision of an accurate result with interpretation for the clinician if required. The output for the clinician is being able to answer the original question. So, if the ECG findings indicating no ST elevation but an increased serum cardiac marker result (e.g., troponin I), the answer is a decision to treat the patient with platelet aggregation inhibitors, in the expectation of a reduced risk of death resulting from further cardiac events. The secondary healthcare provider output is the satisfactory triage of the patient through the system, with minimal risk, optimal use of resources, and discharge to primary care with a good prognosis. The purchaser output is similar to the provider's, with an emphasis on utilization of resources balanced against morbidity and mortality statistics.

In contrast, "silo thinking" or "silo management" encourages the laboratory to focus solely on producing an accurate result, and the clinician in the emergency department to focus on obtaining the result in order to triage the patient to the intervention modality indicated by the findings in the phase of care connected with the stay in the emergency department. In silo thinking, the cardiologist will focus on the intervention and the prognosis, ensuring that any subsequent follow-up addresses risk factors, e.g., hypercholesterolemia and the use of statins. The patients' primary care physician will have to be informed and alerted to any need for monitoring. The provider manager may focus on the patient journey through the organization—but a journey viewed only through the outputs of the departments or silos involved with that specific patient. The purchaser may take a similar perspective. The patient, on the other hand, focuses on the output from the whole cascade of processes set in motion as a result of the initial emergency, namely, whether he will live or die.

If we then consider the case of the patient with chest pain and the initial triage stage in which the cardiac markers were requested (see Figure 3-5b), the use of POCT may not necessarily involve the laboratory in the immediate event, though the rapid result may enable faster intervention with a better clinical outcome (e.g., prognosis and reduced risk of further cardiac events), a shorter stay in the emergency department, and possibly a reduced overall length of stay. This scenario addresses the outcomes of interest to all stakeholders and should enable management to make decisions involving all stakeholders in a coordinated fashion, facilitating the necessary process change.

Evidence-Based Practice

The increasing, or perhaps more explicit, emphasis on outcomes has resulted from the acceptance that clinical practice should be evidence based, encouraging the adoption of the culture of evidence-based medicine (EBM). This culture requires that the best evidence is used in the care of individual patients. When making clinical decisions about individual patients, the evidence must be balanced in terms of benefits and risks, inconvenience, and costs of alternative strategies (3,4). The same arguments apply for evidence-based laboratory medicine (EBLM), where evidence is used to guide decision making in all aspects of diagnostic medicine (5), and thus also for POCT. Thus EBM and EBLM are tools that can help formulate healthcare policy (6) and support the constant drive for improved quality, as well as help make decisions on resource allocation.

The principles of EBLM impact a patient-centered approach to healthcare in two major ways: (1) they can help to provide powerful data on the use of POCT, and (2) they can guide expectations on the outcomes that can be delivered and the changes required in clinical practice to deliver those outcomes. In particular, EBLM principles can serve the following purposes:

- Identifying and validating the clinical questions that prompt the use of diagnostic tests, as well as their delivery by POCT.
- Identifying the decisions likely to be made on receipt of a test result, the actions that might be taken, and the outcomes (and associated benefits) that can be achieved resulting from use of the diagnostic service.
- Identifying the appropriate use of POCT, and the specific benefits achieved from using this testing modality.
- Producing guidelines on clinical practice.
- Determining the content of clinical protocols and care pathways.
- Establishing the quality specifications for the delivery of a diagnostic service.
- Determining the style of diagnostic service provided (e.g., off-site core laboratory, hospital central laboratory, laboratory network, satellite laboratory, POCT, self-testing).
- Providing a standard against which current diagnostic services, as well as clinical practice, whether laboratory or POCT, can be audited.
- Generating the evidence for a business case for a particular diagnostic service and mode of delivery (e.g., POCT).
- Defining the education and training needs of healthcare professionals.

Outcomes

An improved health outcome can be considered to be the "maximization of benefit and the minimization of risk, in the care of individual patients, at reasonable cost." Thus though outcomes might be defined primarily in terms of patient benefit, they should also be measured in terms of health and cost (7).

The outcome of immediate interest to the patient is the *clinical outcome*. However, there can also be a benefit to the caregiver and to the way in which care is delivered; this is called an *operational outcome*. Financial benefits to the patient, the healthcare provider, the healthcare purchaser, or society at large are called *economic outcomes* (8,9). This delineation of outcomes is helpful in evaluating the way that services are delivered, and can be particularly useful in making a business case for the use of POCT.

Different types of outcome also mean that there are a number of beneficiaries from the use of POCT. These stakeholders are interested not only in the quality of the service, but also in the outcomes achieved. It is important to realize that the basic diagnostic performance of a test will invariably have been established using a laboratory-based service, probably in a research setting, and thus the focus of any outcome study on POCT should be directed toward assessing the impact of the alternative testing modality.

Clinical outcomes

The hard or objective measures of clinical outcome are indices of morbidity and mortality. However there are problems with using these indices to validate the use of a test, or a testing modality such as POCT. The most pressing of these is the time that may be required to collect the relevant data. In addition, mortality and morbidity endpoints depend on many elements of the care pathway, including interventions after the provision of a test result (2,9–12). Thus it can be difficult to design studies to show the impact of diagnostic tests on patient outcomes.

To overcome some of these problems, one needs to consider the use of surrogate outcome measures. These measures allow one to study the unique impact of the test result and avoid bias (or adverse influence) that could be introduced by another element of the care pathway. Surrogate markers should have a direct relationship to morbidity and mortality indices. Morbidity represents aspects of the care pathway, e.g., ensuring compliance with therapy, and reducing hospital stay (to minimize the risk of infection). Surrogate markers reflect these aspects (see Figure 4-1, which shows that several of these indices are used in routine clinical practice to monitor the effectiveness of the care pathway such as HbA1c). Consequently the translation of data from an outcome study to a clinical guideline and definition of a care pathway can be additional benefits (or outputs) from an outcome study.

Operational outcomes

Operational outcomes are generally considered part of the economic outcomes. The reason for highlighting them specifically is to identify ways in which adoption of a new technology, in this case POCT, requires a change in practice in order to deliver an improved outcome. In a patient-centered "systems approach to healthcare," the outcomes from the diagnostic element of a care pathway may be viewed as benefits in themselves, but they will also facilitate a benefit (possibly clinical) later in the pathway as a consequence of a change in practice. An example is the use of POCT to reduce "time to treatment." This is an operational benefit in its own right, but the earlier treatment might also improve the patient's prognosis. POCT helps achieve many operational outcomes: reduction in the length of stay and the number of clinic visits, transfer of care to the primary sector, and the use of diagnostic and treatment centers. These changes in practice improve the quality of care for the caregiver and the patient.

Economic outcomes

When considering an economic outcome, it is important to establish which stakeholders will benefit. Pragmatically speaking, if a change in practice is required to generate an economic benefit (for example, reduced costs), then it makes sense to get the perspective of the manager or budget holder responsible for resource reallocation. However, one can also view economics from the perspective of the patient, the caregiver, the provider unit manager (e.g., the emergency department), the provider organization manager, the purchaser organization, or the government (which represents society at large).

Economic outcomes can be viewed a number of ways. For example, one can begin with the cost of the test, then progress through the cost per episode (e.g., the clinic visit), the cost per completed episode (e.g., the removal of malignant parathyroid gland tissue), or the cost of a year of life gained (13). There is a broad link between this hierarchy and the accepted ways of looking at the health economics of interventions such as diagnostic testing: from cost minimization through cost benefit, cost utility to cost effectiveness analysis. Cost minimization is not worth further consideration in relation to POCT, because it focuses solely on the cost per test, which is the province of the "testing budget silo." This is quite inappropriate when trying to look at the benefits of different care modalities through the use of different testing modalities. Also, the cost of the POCT device (reagent) is likely to be greater, because of the economies of scale that a central laboratory is more likely to achieve. However, if POCT saves a clinic visit, then the cost associated with clinic time is reduced.

A number of surrogate outcome measures have an important place in the analysis of economic outcomes related to the use of diagnostic tests, particularly in the case of POCT (Figure 4-1). Specifically, they enable assessment of short- and long-term costs and benefits. At a local decision-making level, for example, shorter-term gains are more likely to affect the resource reallocations that accompany the change in practice, thereby leveraging or achieving the benefit. Longer-term gains are of greater value in governmental health policymaking.

Obtaining and Reviewing the Evidence

The quality of evidence on the use of diagnostic tests is poor, both in terms of the information that should be included in any study as well as the design of the study itself (14–19). To overcome the issue of poor reporting of study results, many key journals have accepted the STARD guidelines (20,21). These guidelines provide a good template for designing a robust study, identifying the correct data to collect, clarifying how to report the completed study, reviewing published studies, and assessing the quality of studies when performing a systematic review.

Obtaining evidence

A good study protocol depends upon a careful consideration of the "question-test-decision-action cascade" and identifying

- an explicit clinical question that is prompting the request for a test,
- the decision to be made upon receipt of the test result,
- the action (treatment or other intervention) to be taken, and
- the expected outcome from the action or intervention.

The following points must be considered:

- An explicit recognition of the patient group and the setting in which patients will be studied.
- A clear statement of the clinical question.
- A clear understanding of the decision that would be made upon receipt of the test result.
- Identification of the action(s) that would be taken, based on the decision being made.
- Identification of an expected outcome from the action being taken.
- An understanding of any confounding factors that might lead to interrupting the "clinical question to action" cascade.

High-quality diagnostic studies have a number of key features (Table 4-1). Rainey (10) has suggested that such features can provide a template for designing "manageable and meaningful" outcome studies. Addressing study design for POCT, Rainey proposed an additional set of principles specifically for outcome-based studies:

- Use a prospective study, ideally with a randomized or crossover design.
- Minimize sources of variability, i.e., limit the study to a single service, diagnosis, or procedure.
- Ensure consistent interventions in response to test results.
- Choose outcome measures that can be quantified.
- Use outcome variables that are determined in close temporal proximity to the testing process.
- Use outcome measures that are good surrogates for long-term outcome measures (for example, HbA1c, cholesterol, urine albumin–creatinine ratio).
- Use outcome measures that have intrinsic financial value, such as blood product usage or length of stay.
- Identify local factors that might impact the results and therefore limit the wider application of the data.

When reporting studies, include the key features from the STARD guidelines (these can be summarized using the criteria listed in Table 4-1) (20,21).

TABLE 4-1. A short checklist of features that should be found in an outcomes study associated with Point-of-Care Testing

- Clear statement of research question—namely, specific use of POCT.
- Identification of study population, including inclusion and exclusion criteria.
- Number of patients studied.
- Identification of characteristics of patients recruited, e.g., symptoms, severity of disease.
- Identification of how patients were recruited, e.g., prospective or consecutive.
- Identification of trial design.
- Identification of decision made and action taken upon receipt of result.
- Identification of outcome measure used.
- Description of all methods used and their performance characteristics.
- Identification of whether operators were blinded to outcome measure.
- Documentation of results including a "two by two" table where applicable.
- Description of statistical methods used.
- Indication that attempts were made to ensure change of practice where required to deliver improved outcome.
- Indication of how many patients completed study.
- Documentation of confounding factors noted during study.
- Discussion of data and applicability of results.

It is generally accepted that the randomized controlled trial (RCT) best evaluates the effectiveness of an intervention, because it this minimizes bias and the impact of confounding variables (19,22,23). However, the randomized design was originally conceived for the trial of a therapeutic intervention, and is not always considered the best approach for a diagnostic test.

With respect to POCT, an RCT could involve randomizing patients to the usual testing approach (i.e., the central laboratory) or to POCT (19). Ideally the trial should not only examine the testing process but also the steps afterwards—namely the decisions and outcomes that follow—but that is not always possible. Cagliero et al. (24) compared POCT testing against the central laboratory approach to assess the value of POCT for HbA1c in a diabetes clinic. They demonstrated a greater decrease in HbA1C in the POCT patients compared to those who received their results later from the central laboratory. These authors were effectively looking at the impact of immediate patient counseling vs. counseling undertaken when the laboratory result was received later, after the patient had departed. This study is not therefore a trial of POCT itself but rather of the whole diagnostic and decision process. This is an important point because it highlights change in a care process of which POCT is only one, albeit a key, element.

An RCT enables the unique contribution of the test result to be assessed in terms of its impact on decision making, because one is looking at an outcome that requires making a decision immediately after the result is available.

A retrospective analysis will only show the relationship between a result and the outcome; it will not show whether the result influenced the clinical decision or the outcome.

However, an RCT also has some problems, primarily because the test result is only one part of a decision-making process that also requires understanding and action upon receipt of the result. "Understanding" may include "education" and ongoing "counseling" of the patient, such as that required with management of chronic conditions like diabetes and anticoagulation therapy. Other difficulties include controlling POCT "confounders," such as when the result is not accessed or used, or where the local clinical practice has not been adjusted to take into account the rapid receipt of the result. Kilpatrick and Holding (25), for example, found that over 40% of results electronically transmitted to the emergency department, in this case from a laboratory, were never accessed (and therefore not acted upon). Kendall et al., who conducted a randomized controlled trial of POCT in the emergency department, found no reduction in the length of stay, despite a significant reduction in the time taken to deliver the result (26). The authors concluded that there may not have been a vital change in practice, such as staff accessing the results as soon as they became available. Nichols et al., had a similar experience using POCT in an interventional radiology unit; the study showed that improved outcomes were only delivered when changes in practice were instituted (27). Thus, confounders in the study design can become the key pointers to change in practice at the time of implementation.

There are few RCTs of diagnostic tests, let alone of POCT, and therefore other study designs have been employed to gather needed evidence. Lijmer and others (18,28,29) reviewed studies on diagnostic tests, including outcomes studies, and found that some forms of study design can lead to significant bias in the results. The most common form of bias comes from case-controlled study designs in which the study and control populations are different (e.g., comparing patients diagnosed as having the disease to those who do not). In reality, the test will be used in patients with unclear or borderline symptoms, and in this situation, test performance will be different and inferior compared to its performance under the ideal situations in which it was assessed. Thus measures of test quality, such as the likelihood ratio, will depend on which patients are studied; this form of bias is sometimes called selection or spectrum bias. Sackett and Haines described this bias well in relation to the use of brain natriuretic peptide (BNP). They stated the importance of assessing any test in a cohort of truly consecutive patients presenting in the same way as in the setting where the test would be used (30).

Case-control studies and spectrum bias are the most common problems of many diagnostic studies; however, they are only two among many others. One example is the use of a poor reference test or "gold standard" with which to compare the test or protocol under investigation. This practice results in underestimating test accuracy, depending upon the prevalence of the target condition being investigated (31). Another example is "review bias," which occurs when the interpretation of the test results is affected by knowledge of the reference test, or vice versa. This may be significant if the test method assessment is subjective, such as in the visual reading of urinary dipsticks (32). A third example is "verification bias," of which there are several types. All cases of verification bias involve not applying the same reference test or protocol to all the subjects who underwent the index test or protocol (28,33).

Bossuyt describes three types of study design for studies of diagnostic accuracy: (1) the randomized studies described previously; (2) paired studies; and (3) before and after studies (34). The paired design involves undertaking both tests in the same patient, which can reduce one form of variability. However it is not possible to use this in an outcomes study where the result is acted upon and in which the outcome is influenced by the test result. Therefore, it would not be appropriate for assessing the utility of POCT. The "before and after" study is a special form of the paired design (35). In this approach, the outcome indices associated with the routine approach to care are collected and then, after a period of training, a POCT strategy is adopted and outcome indices collected at the appropriate time. The two sets of indices are then compared. This approach has been used for a number of studies of POCT in the primary care setting and may be of more practical relevance. However, the investigator must ensure that all other aspects of the care pathway are the same in the "before" and "after" phases of the study. In Rink et al.'s "before and after" study in primary care, the "before" and "after" phases were then followed by another phase of "before" care (36). Another way to look at the "before and after" study is that it investigates the "test-to-treat" element of the care pathway.

All of the above study designs carry the risk of the Hawthorne effect, i.e., that patients and the caregivers may be aware that a study is taking place and therefore may take extra care, causing bias and improving the potential outcomes. A cross-sectional study can serve as an alternative approach; this approach is thought to provide evidence of inferior quality, because other factors could affect the outcome beyond that exclusively due to the test result. However, it is difficult to determine whether providing a POCT result causes any achieved outcome in an observational study. Though this is understandable, it could be argued that the observational study can provide a more holistic picture of the care process. Thus a meta-analysis by Coster et al. (37) that focused on RCTs suggested no evidence to support the use of frequent self-monitoring of blood glucose; however, Karter et al. (38) and Schiel et al. (39) showed that in studies of large populations of diabetics participating in a diabetes education and management program, those who tested themselves more frequently had lower HbA1c results. These studies did not identify frequent testing as the unique feature that reduced HbA1c, but they did show that a pattern of care that included more frequent testing did reduce HbA1c. Note that in any observational or cross-sectional study, there are greater risks of differences in the populations or the style of care influencing the final outcome.

Reviewing evidence

Statistical analysis of data is important. Several good texts show the required techniques (40,41). Much of the data on diagnostic testing has been based on data obtained from the numbers of true and false, positive and negative observations (the so called "two by two" table). This table enables generating data on sensitivity and specificity and constructing receiver operator curves. However, as Deeks (42) pointed out, the receiver operator curve is not easy to interpret in a routine clinical situation. Instead, the likelihood ratio is more meaningful in this context, especially when it is connected to the Fagan nomogram, which links the pre- and post-test probability to the likelihood ratio observed (43). In this way the clinician is able to quickly (1) identify whether a test result is helpful in "ruling in" or "ruling out" a particular course of action and (2) see how the change in the likelihood ratio and, therefore, the probability of having (or not having) a disease is related to a particular result.

Having a number of studies available on a particular topic such as POCT enables a meta-analysis of the data (44) to be performed. The benefit is that the variability of the data among different studies can be assessed, which increases the confidence in the data, especially when the variability (or heterogeneity) of the data is low. The construction of a "Forest plot" is a good visual means of assessing the variability, as well as the overall quality, of the data (45). Pooling data in this way can also indicate when the addition of further data is unlikely to improve its quality, thereby implying that no further studies need to be undertaken.

If evidence is to be reviewed for possible inclusion in a systematic review and meta-analysis, then the investigator must determine the quality of each study and whether it should be included or excluded from the review. These measures of quality are largely based on the criteria for study design and reporting. A number of publications have set scoring criteria for determining which papers to include in a review. These criteria thus allow papers, and therefore the quality of the data they contain, to be assessed objectively (46,47), which is extremely important when considering meta-analysis. In many papers there may be a wealth of references extracted from the literature search, but few included in any formal meta-analysis. This is typically because there are differences in either the patient population or in the intervention. An example of the former is the meta-analysis of the use of the random urine protein-creatinine ratio, when cumulative analysis could only be performed on patients with preeclampsia and with a similar pre-test probability (48). An example of the latter is the very small proportion of papers that could be included in a meta-analysis describing RCTs of self-monitoring of blood glucose, because of the variation in the interventions reported in the literature (49,50).

A systematic review is a good source of robust information and brings together a number of studies which in themselves may not have been sufficient to give confidence on the utility of a test. When meta-analysis is possible, it will provide pooled data; a cumulative meta-analysis can indicate the significance of the findings as studies are added to the analysis. If there is no further change in the significance of the results, then the best utility of the test has likely been achieved, and no further studies are warranted. The variation between studies (called "heterogeneity") will indicate the robustness of the study data and the existence of confounding variables. This can be helpful in designing further studies, as well as in introducing the test into routine practice.

PROCESS CHANGE

Process change is one of the most challenging aspects to the introduction of POCT, and it has two key components (1) changing the elements of delivery in the care pathway and (2) reallocating resources (1). The challenges are threefold:

1. POCT requires caregivers to take on new tasks and practices, for example, performing diagnostic tests and undertaking quality control procedures.
2. POCT may require a number of healthcare professionals to change the way they perform existing processes of care.
3. POCT invariably requires some reallocation of finances—perhaps the most delicate issue of all.

Yet without these changes there is no point in implementing POCT—it is part of the incentive to change.

Much has been written about implementing change in business processes. Some of these changes have been applied to aspects of healthcare. Thus LEAN management principles (made famous by Toyota) have been applied to the healthcare environment, with the objective of improving efficiency and outcomes (51). Process improvement using Six Sigma principles has also been applied in the healthcare environment (52,53). However, there is only a limited amount to be gained, and perhaps something to be lost, from applying these techniques to only one facet (or "silo") of patient care. In cases where these techniques have been applied to a clinical setting, benefits have been identified, as have the difficulties of changing culture and working practices (54). The challenge, therefore, is to apply these principles to the whole patient care pathway, and in relation to diagnostic testing and specifically POCT, beyond the "testing phase."

Changing the Care Pathway

The starting point for changing process in the clinical setting is a thorough analysis of the current care pathway, the related working practices, and then the confounding influences. Three examples (below) show the key elements of the analysis and highlight the changes required to deliver the improved outcome.

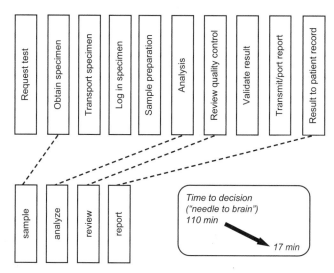

FIGURE 4-2. Schematic representation of the care pathway showing the reduction in the steps required to obtain a cardiac marker result using POCT in the case of a patient admitted to the emergency department with chest pain, and the concomitant reduction in time needed to make a decision on triage of the patient (based on the findings of Lee-Lewandrowski et al. [55]).

In this first example, POCT testing is used to *facilitate rapid delivery of results* for a patient admitted to the emergency department (ED) with acute chest pain. Studies demonstrate that POCT can help to reduce length of stay in the ED (55–57) and reduce the number of patients admitted to the coronary care unit, presumably in part by improving the accuracy of diagnosis and therefore leading to more appropriate patient triage. Figure 4-2 illustrates the process of care around the request for cardiac markers on admission using a laboratory service, together with the changes that would be expected from implementing POCT. This scenario would be similar for any situation in which a diagnostic test is used to influence a decision.

The second example is more indicative of *long-term disease management* where studies have shown a benefit from having the laboratory results (that give a measure of current disease status) available at the time of the clinic visit, for example, in the case of diabetes. Figure 4-3 illustrates the process around the request for biochemical markers such as HbA1c, cholesterol, and urine albumin-creatinine ratio using a laboratory service, together with the changes that would be expected from implementing POCT. The scenario would be similar for other clinical review visits associated with long-term disease management, as well as for a first visit to a primary care physician with a fairly simple presentation requiring a test to "rule in" or "rule out" the disease (e.g., urinary tract infection, sore throat, influenza).

The third scenario is slightly more speculative—*a patient diagnosed with a condition that requires pharmaceutical intervention*—but nonetheless valid,

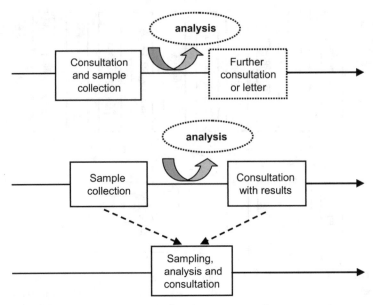

FIGURE 4-3. Schematic representation showing the care pathway for a patient with diabetes mellitus attending a clinic for regular review. In the two uppermost options, the patient either attends for the consultation and blood-drawing for HbA1c, with the result being communicated at another visit or by telephone or letter, or the patient attends before the consultation to have the blood drawn, the result being available at the consultation. The third option, facilitated by POCT, enables the blood to be taken, the analysis performed, and the consultation informed by the result all taking place at one visit.

as the objective of each of these exemplars is to prompt the manager or policymaker to look for the potential changes in the care process. A test result may be required to assess the required drug dose, perhaps as part of a prognostic assessment, or to determine which drug is to be used—an illustration of the concept of "personalized medicine," sometimes called "theranostics," which is giving the right drug, in the right amount, to the right patient, at the right time (58). If a patient with psoriasis is placed on azothioprine, then it is now accepted practice to assess the patient's thiopurine methyltransferase (TPMT) status to learn the rate at which the drug is likely to be metabolized; an excessive dosage of drug, possibly as a result of slow metabolism, could lead to fatal bone marrow damage (59). Of course it would be possible to monitor the patients with other tests for early signs of drug toxicity, but that involves additional tests, and additional clinic visits. Figure 4-4 illustrates the process around the request for a measurement of plasma TPMT status using a laboratory service, together with the changes that would be expected from the implementation of POCT. An alternative would be the use of BNP to assess the degree of heart failure and to guide therapy; in this case the follow-up visits to assess the efficacy of treatment and to modify the treatment based on the change in plasma BNP levels would also optimize therapy (60).

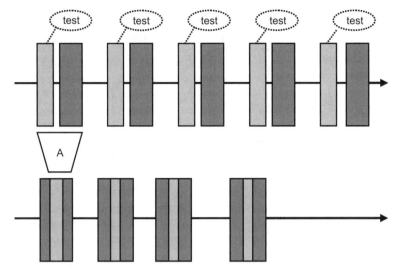

FIGURE 4-4. Schematic representation of a care pathway in which a patient attending a clinic requires a test for determining the dosage of a drug, with subsequent visits to assess efficacy of dose and optimization of therapy. In the upper scenario, using a laboratory service, two visits are required: (1) to provide a blood sample and (2) to provide consultation. In the lower scenario, using POCT, we see fewer visits and faster optimization. The scenario of the first visit (A) could also reflect the case of a patient requiring a test for drug candidacy, e.g., Her-2/neu status and Herceptin, or to assess dosage, e.g., TPMT status and azothioprine. Key: the light shading denotes the phlebotomy visit, and the dark shading, the consultation visit.

Reallocating Resources

This aspect of management will vary by country and depend on its approach to allocation of resources or reimbursement. As noted in Chapter 3, when taking a systems approach to managing a business, the manager must know in detail the workings of all steps in the process, and how they relate to each other. This invariably will involve budgetary silos, and the skill is to be able to control those silos in a way that prevents the silo management process from impeding financial flows between silos within the framework of a care pathway.

A care pathway or a completed patient episode typically involves a contribution from a number of budgetary groups within the organization. The nature of the budgetary groups may vary slightly between organizations but generally includes clinical staff, nursing staff, operating rooms, laboratory medicine, radiology, clinical support services, drugs and dressings, facilities (wards, clinics, rooms), and overheads. To illustrate the nature of the financial transactions involved in the reallocation of resources associated with the introduction of POCT, we will examine testing for coagulation status during open heart surgery. A number of studies have shown that POCT in this case reduces the requirement for transfused blood products and the length of post-operative

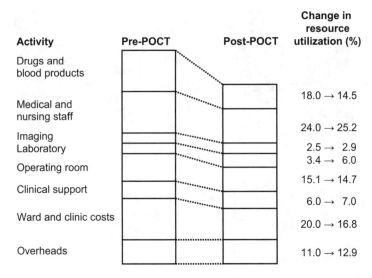

FIGURE 4-5. An illustration of the change in resource utilization associated with cardiac surgical procedures and the impact of intra-operative assessment of coagulation status. The figures are based on average costs and using 1998 prices.

stay (61). Figure 4-5 illustrates the changes in the average cost associated with cardiac surgery before and after the introduction of POCT.

The increased investment in POCT is offset by the reduction in blood products and in bed stay, resulting in an overall net gain. Chapter 5 presents further examples of these net gains, including the impact of good glycemic control on the costs of diabetes care, where the main benefits come from reduced emergency admissions and hospital stay, and the use of POCT in the emergency department, which ultimately reduces length of hospital stay. All of these examples demonstrate that knowledge of the financial allocations for each aspect of the elements of care, and their interrelationships, are crucial to good financial and resource management to improve overall health outcomes.

VALUE ADDED

In business terms, a process "adds value" if it meets three criteria: (1) the customer is willing to pay for the process or activity, (2) the process or activity physically changes or transforms a product or service, and (3) the process or activity is performed correctly on the first attempt. A process or activity does not add value if it involves preparation or set up, if it is focused on control or inspection, if it simply results in moving a product from one place to another, or if the process leads to delays or failure (1).

It is possible to translate this business perspective into a healthcare setting, and it is clear that the greatest added value from POCT comes from applications where clinical outcomes improve—and where a value can then be ascribed to individual patients, to their families, to their employers, and to society as a whole. However there is also the potential for added value to the healthcare provider (reduced bed requirements, reduced cost of care, reduced readmissions, enhanced reputation), as well as to the purchaser (perception of good reputation, high quality of care reflected in health statistics, reduced cost of care).

It is suggested that when analyzing the value-added element of any process, sub-processes should undergo a detailed analysis to determine which ones are adding the value. This is, in fact, part of the approach to conducting trials of new interventions such as POCT. It is important to demonstrate that the value-adding elements of a process can be identified, because this is the only way that investments in new technology can be managed responsibly. Identifying the added value provides the incentive for investment and change. Focusing on the cost per test, or any financial metric other than that associated with an episode of care (which may be a year of life gained) will not provide the answer required to determine whether POCT will be beneficial in terms of health outcomes.

SUMMARY AND RECOMMENDATIONS

Understanding the care pathway from presentation to outcome helps to improve health outcomes. In the case of POCT this entails the following:

- Knowing the clinical question being asked when a test is requested.
- Understanding the "question-test-decision-action" cascade.
- Recognizing all stakeholders involved in the appropriate use of a diagnostic test.
- Being able to evaluate the evidence to demonstrate the clinical utility of a test, including its delivery through POCT, and the impact on outcomes in the design, execution, reporting, and review of studies.
- Being able to recognize the added value.
- Being able to define the process change required to maximize the benefits for all stakeholders, as well as the steps required to make that change.
- Translating the outcomes into specific outputs applicable to each of the stakeholders.

REFERENCES

1. Harmon P. Business process change: a manager's guide to improving, redesigning, and automating processes. Amsterdam: Morgan Kaufmann Publishers, 2003:528pp.
2. Sloan FA, ed. Valuing health care: cost, benefits, and effectiveness of pharmaceuticals and other medical technologies. Cambridge: Cambridge University Press, 1998:273pp.
3. Sackett DL, Haynes RB, Guyatt GH, Tugwell P. Clinical epidemiology: a basic science for clinical medicine, 2nd ed. Toronto: Little, Brown, 1991:441pp.

4. Guyatt G, Haynes B, Jaeschke R, et al. Introduction: the philosophy of evidence-based medicine. In: Guyatt G, Rennie D, eds. User's guide to the medical literature: a manual for evidence-based clinical practice. Chicago: American Medical Association (JAMA and Archive Journals), 2002:3–12.

5. Price CP. Benchmarking in laboratory medicine: are we measuring the right outcomes? Benchmarking 2005;12:449–66.

6. Muir Gray JA. Evidence-based healthcare: how to make health policy and management decisions. Edinburgh: Churchill Livingstone, 1997:270pp.

7. Bissell MG, ed. Laboratory-related measures of patient outcomes: an introduction. Washington, DC: AACC Press, 2000:193pp.

8. Price CP. Evidence-based laboratory medicine: supporting decision-making. Clin Chem 2000;46:1041–50.

9. Price CP. Application of the principles of evidence-based medicine to laboratory medicine. Clin Chim Acta 2003;333:147–54.

10. Rainey PM. Outcomes assessment for point-of-care testing. Clin Chem 1998;44:1595–6.

11. Price CP. Point-of-care testing: potential for tracking disease management outcomes. Dis Manage Health Outcomes 2002;10:749–61.

12. St John A, Price CP. Measures of outcome. In: Price CP, Christenson RH, eds. Evidence-based laboratory medicine: from principles to outcomes. Washington, DC: AACC Press, 2003:55–74.

13. Marshall DA, O'Brien BJ. Economic evaluation of diagnostic tests. In: Price CP, Christenson RH, eds. Evidence-based laboratory medicine: from principles to outcomes. Washington, DC: AACC Press, 2003:159–86.

14. Reid MC, Lachs MS, Feinstein AR. Use of methodological standards in diagnostic test research: getting better but still not good. JAMA 1995;274:645–651.

15. Mulrow CD, Linn WD, Gaul MK, Pugh JA. Assessing quality of a diagnostic test evaluation. J Gen Intern Med 1989;4:288–95.

16. Jaeschke R, Guyatt G, Sackett DL. Users' guides to the medical literature. III. How to use an article about a diagnostic test. A. Are the results of the study valid? Evidence-Based Medicine Working Group. JAMA 1994;271:389–91

17. Jaeschke R, Guyatt GH, Sackett DL. Users' guides to the medical literature. III. How to use an article about a diagnostic test. B. What are the results and will they help me in caring for my patients? The Evidence-Based Medicine Working Group. JAMA 1994;271:703–7.

18. Lijmer JG, Mol BW, Heisterkamp S, et al. Empirical evidence of design-related bias in studies of diagnostic tests. JAMA 1999;282:1061–6.

19. Bossuyt PM, Lijmer JG, Mol BW. Randomized comparisons of medical tests: sometimes invalid, not always efficient. Lancet 2000;356:1844–7.

20. Bossuyt PM, Reitsma JB, Bruns DE, et al. Towards complete and accurate reporting of studies of diagnostic accuracy: the STARD initiative. Standards for Reporting of Diagnostic Accuracy. Clin Chem 2003;49:1–6.

21. Bossuyt PM, Reitsma JB, Bruns DE, et al. The STARD statement for reporting studies of diagnostic accuracy: explanation and elaboration. Clin Chem 2003;49:7–18.

22. Moher D, Schulz KF, Altman DG for the CONSORT group. The CONSORT statement: revised recommendations for improving the quality of reports of parallel group randomized trials. JAMA 2001;285:1987–91.

23. Bruns DE, Huth EJ, Magid E, Young DS. Toward a checklist for reporting of studies of diagnostic accuracy of medical tests. Clin Chem 2000;46:893–5.

24. Cagliero E, Levina E, Nathan D. Immediate feedback of HbA1c levels improves glycemic control in type 1 and insulin-treated type 2 diabetic patients. Diabetes Care 1999;22:1785–9.

25. Kilpatrick ES, Holding S. Use of computer terminals on wards to access emergency test results: a retrospective audit. BMJ 2001;322:1101–3.

26. Kendall J, Reeves B, Clancy M. Point-of-care testing: randomised, controlled trial of clinical outcome. BMJ 1998;316:1052–7.

27. Nichols JH, Kickler TS, Dyer KL, et al. Clinical outcomes of point-of-care testing in the interventional radiology and invasive cardiology setting. Clin Chem 2000;46:543–50.

28. Begg CB. Biases in the assessment of diagnostic tests. Stat Med 1987;6:411–23.

29. Mower WR. Evaluating bias and variability in diagnostic test reports. Ann Emerg Med 1999;33:85–91.

30. Sackett DL, Haynes RB. The architecture of diagnostic research. BMJ 2002;324:539–41.

31. Irwig L, Bossuyt P, Glasziou P, et al. Designing studies to ensure that estimates of test accuracy are transferable. BMJ 2002;324:669–71.

32. Moons KG, Grobbee DE. When should we remain blind and when should our eyes remain open in diagnostic studies? J Clin Epidemiol 2002;55:633–6.

33. Knottnerus JA. The effects of disease verification and referral on the relationship between symptoms and disease. Med Dec Making 1987;7:139–48.

34. Bossuyt PM. Study design and quality of evidence. In: Price CP, Christenson RH, eds. Evidence-based laboratory medicine: from principles to outcomes. Washington, DC: AACC Press, 2003:75–92.

35. Knottnerus J, Dinant G-J, van Schayk O. The diagnostic before-after study to assess clinical impact. In: Knottnerus J, ed. The evidence base of clinical diagnosis. London: BMJ Books, 2002:81–93.

36. Rink E, Hilton S, Szczepura A, et al. Impact of introducing near patient testing for standard investigations in general practice. BMJ 1993;307:775–8.

37. Coster S, Gulliford MC, Seed PT, et al. Monitoring blood glucose in diabetes mellitus: a systematic review. Health Tech Assess 2000;4:1–93.

38. Karter AJ, Ackerson LM, Darbinian JA, et al. Self-monitoring of blood glucose levels and glycemic control: the Northern California Kaiser Permanente Diabetes Registry. Am J Med 2001;111:1–9.

39. Schiel R, Muller UA, Rauchfub J, et al. Blood-glucose self-monitoring in insulin-treated Type 2 diabetes mellitus: a cross-sectional study with an intervention group. Diabet Metab 1999;25:334–40.

40. Altman DG, Machin D, Bryant TN, Gardner MJ, eds. Statistics with confidence, 2nd ed. London: BMJ Books, 2001:240pp.

41. Egger M, Davey Smith G, Altman DG, eds. Systematic reviews in health care; meta-analysis in context. London: BMJ Books, 2001:487pp.

42. Deeks JJ. Systematic reviews in healthcare: Systematic reviews of evaluations of diagnostic and screening tests. BMJ 2001;323:157–62.

43. Fagan TJ. Nomogram for Bayes theorem. NEJM 1975;293:133–4.

44. Irwig L, Macaskill P, Glasziou P, Fahey M. Meta-analytic methods for diagnostic test accuracy. J Clin Epidemiol 1995;48:119–30.

45. Boyd JC, Deeks JJ. Analysis and presentation of data. In: Price CP, Christenson RH, eds. Evidence-based laboratory medicine: from principles to outcomes. Washington, DC: AACC Press, 2003:115–36.

46. Downs SH, Black N. The feasibility of creating a checklist for the assessment of the methodological quality both of randomised and non-randomised studies of healthcare interventions. J Epidemiol Community Health 1998;52:377–84.

47. Whiting P, Rutjes AW, Dinnes J, et al. Development and validation of methods for assessing the quality of diagnostic accuracy studies. Health Technol Assess 2004;8:iii,1–234.

48. Price CP, Newall RG, Boyd JC. Use of protein:creatinine ratio measurements on random urine samples for prediction of significant proteinuria: a systematic review. Clin Chem 2005;51:1577–86.

49. Welschen LM, Bloemendal E, Nijpels G, et al. Self-monitoring of blood glucose in patients with type 2 diabetes who are not using insulin: a systematic review. Diabetes Care 2005;28:1510–7.

50. Sarol JN Jr, Nicodemus NA Jr, Tan KM, Grava MB. Self-monitoring of blood glucose as part of a multi-component therapy among non-insulin requiring type 2 diabetes patients: a meta-analysis (1996–2004). Curr Med Res Opin 2005;21:173–84.

51. Bahensky JA, Roe J, Bolton R. Lean sigma-will it work for healthcare? J Healthc Inf Manag 2005;19:39–44.

52. Woodard TD. Addressing variation in hospital quality: is Six Sigma the answer? J Healthc Manag 2005;50:226–36.

53. Guinane CS, Davis NH. The science of Six Sigma in hospitals. Am Heart Hosp J 2004;2:42–8.

54. Christianson JB, Warrick LH, Howard R, Vollum J. Deploying Six Sigma in a healthcare system as a work in progress. Jt Comm J Qual Patient Saf 2005;31:603–13.

55. Lee-Lewandrowski E, Corboy D, Lewandrowski K, et al. Implementation of a point-of-care satellite laboratory in the emergency department of an academic medical center. Impact on test turnaround time and patient emergency department length of stay. Arch Pathol Lab Med 2003;127:456–60.

56. Ng SM, Krishnaswamy P, Morissey R, et al. Ninety-minute accelerated critical pathway for chest pain evaluation. Am J Cardiol 2001;88:611–7.

57. McCord J, Nowak RM, McCullough PA, et al. Ninety-minute exclusion of acute mycardial infarction by use of quantitative point-of-care testing of myoglobin and troponin I. Circulation 2001;104:1483–8.

58. Sadee W, Dai Z. Pharmacogenetics/genomics and personalized medicine. Hum Mol Genet 2005 Oct 15;14 Spec No. 2:R207–14.

59. Duley JA, Florin TH. Thiopurine therapies: problems, complexities, and progress with monitoring thioguanine nucleotides. Ther Drug Monit 2005;27:647–54.

60. Richards AM, Troughton R, Lainchbury J, Doughty R, Wright S. Guiding and monitoring of heart failure therapy with NT-ProBNP: concepts and clinical studies. J Card Fail 2005;11(5 Suppl):S34–7.

61. Despotis GJ, Joist JH, Goodnough LT. Monitoring of hemostasis in cardiac surgical patients: impact of point-of-care testing on blood loss and transfusion outcomes. Clin Chem 1997;43:1684–96.

Chapter 5

Case Studies in Point-of-Care Testing: Evidence, Benefits, and Challenges

The potential benefits from the use of point-of-care testing (POCT) arise from the need to make a clinical decision and institute therapy or some other intervention at the time the health professional sees or interacts with the patient. The objective is to improve the health outcome, which can be defined as maximizing the benefit, while minimizing the risk in the care of individual patients, at reasonable cost (1).

The clinical need can be defined in a number of ways, and there will be a corresponding range of potential benefits; yet there is a lack of, and a desire for, a broader range of outcomes data to document evidence of those benefits (2). Historically, the earliest clinical decisions aided by POCT involved a simple test or observation that assisted in making a diagnosis—the case in point being tasting urine and recognizing diabetes. As testing technology developed, POCT became associated with acute life-threatening clinical situations such as hyper- or hypoglycemic episodes (in the case of the diabetic), the emergency admission of the unconscious patient, or the management of a patient in intensive care. POCT came to be seen as a key feature of efficiently and effectively managing acutely and critically ill patients. More recently, with its use in clinics, POCT is viewed as a way to optimize a therapeutic intervention and subsequently assess compliance with therapy. In all of these scenarios there is an associated gain for patients, health professionals, and healthcare provider organizations. These gains can be measured in terms of patient and clinician satisfaction as well as in the cost of care. Thus POCT has also evolved as a means of improving the efficiency and the effectiveness of healthcare.

POCT's success in delivering benefits will invariably depend on changing existing clinical and organizational practices (3). Change in any form of organization is difficult—and therein lies the challenge.

This chapter offers illustrative case studies in which POCT has been investigated as a means of improving health outcomes. Not only are the clinical needs defined, the evidence assessed, and the benefits identified—the change in practice required to deliver these benefits is identified, and the management challenges associated with the change in practice are discussed. The cases are not exhaustive, and readers are referred to other texts for a greater insight into each of the clinical scenarios, and for a more comprehensive review of the evidence (4,5).

WHAT TO LOOK FOR IN POCT STUDIES

Evaluating POCT studies can offer instructive information for potential use in your institution or healthcare provider organization. The following questions can help you determine the relevance of each study to your own situations:

- Is the background similar to your area of interest?
- Does the clinical question posed reflect your needs?
- Does the clinical setting reflect your situation?
- Are the patients studied similar to your own?
- Is there evidence available?
- Are there benefits identified?
- Is there a change in practice required?
- Is the change in practice achievable?

Case Study 1: Primary Care—Rapid Diagnosis, Urinary Tract Infection

Background

The suspicion of urinary tract infection (UTI) is a common presentation in primary care. A telephone survey in the USA revealed that 11% of women 18 years or older reported at least one presumed UTI in the last year (6), and, an estimated 60% of women suffer the condition at some stage in their lives (7). Approximately 23% of all hospital-acquired infections are due to UTI (8). The incidence is increased in elderly men and women, particularly among those living in institutions, where the incidence rate can be up to 53% and 37% in women and men, respectively (9).

In certain circumstances and patients, such as pregnant women, diagnosis through bacterial culture is important, because failure to detect a UTI can have serious consequences. However, uncomplicated UTI in the non-pregnant adult female rarely causes severe illness or has significant long-term consequences, and 50% of patients improve without antimicrobials within three days (10). Nevertheless, suspected UTI is one of the most common indications for antimicrobial use, with treatment often based on clinical symptoms and signs without any confirmation by culture (11). Treatment without confirmation is a concern today because of the increase in antimicrobial resistance thought to be caused by inappropriate and unnecessary antimicrobial use (12). Such empirical treatment of women with suspected uncomplicated UTI has been shown in one decision analysis study to be the most cost-effective strategy. However, adding a dipstick test and using the results to "rule out" treatment could be justified if reducing antimicrobial treatment is an additional objective (13). The emergence of multiresistant extended-spectrum beta-lactamase-producing (ESBL) *Escherichia Coli* emphasizes the increasing risks associated with unnecessary antimicrobial use (14,15), which is already known to increase the risk of subsequent UTIs (16).

The frequency of UTI generates a significant workload for the laboratory with large laboratories analyzing 200–300 urine specimens daily (17). Most of the specimens sent to a laboratory will show no evidence of infection when tested, and consequently there has been considerable interest in discovering ways to screen out negative specimens prior to processing them for culture by a "rule out" test strategy (18). This could be particularly appropriate for screening out asymptomatic bacteriuria in a population with low prevalence of UTI, such as pregnant women, and this would be a cost-effective alternative to culture (19,20). However a recent review commissioned by the National Institute for Health and Clinical Excellence (NICE) rejected this advice on the basis of current evidence, though highlighting the problem as an important research topic (21).

Clinical question

Can a random urine dipstick test performed in the primary care physician's clinic for the presence of leukocyte esterase (LE) or nitrite (N) be used to "rule out" the presence of UTI and thereby limit the need for further laboratory investigations?

Setting

Patients presenting in primary care with symptoms of UTI.

Evidence

Systematic reviews and meta-analysis of studies using these tests in adults and children have appeared in recent years and have identified deficiencies in many of the included studies (22–24). Gorelich et al. showed that in children, both the gram stain and dipstick analysis for LE and N performed similarly in detecting UTI in children (used as "rule in" tests) and were superior to microscopic analysis for pyuria (22). A more recent systematic review and meta-analysis in children included a larger number of papers that Gorelich reviewed, and concluded that pyuria ≥ 10/hpf (or μL) and bacteriuria, as detected by microscopy, were best suited for diagnosing UTI in children although there was insufficient data to draw conclusions about the value of urinary dipstick tests when used as a "rule in" test (23).

Three systematic reviews of UTI dipstick data specifically for "ruling out" UTI have been published. One involving adult patients concluded that a positive LE or N test was the best index or "rule-in" test, but a negative dipstick result ("rule out" test) could not exclude UTI in patients with a high prior

probability of contracting the condition (24). Deville et al. (25), using a grad-
ing system to review and evaluate papers, found that the use of nitrite or leukocyte esterase was a valuable "rule out" test; the papers included in their review
had a negative predictive value of 84–98%, the majority being above 95%. St
John et al. (26) used more objective criteria and found similar negative predictive values between 81% and 99%. Interestingly, St John analyzed the studies
included in Deville et al.'s review from the viewpoint of the transition of pre-
test to post-test probability. This suggested that the use of a leukocyte esterase
or nitrite combination in a "rule out" strategy would give an acceptable post-
test probability of 5% or less, especially in the situation where the pre-test
probability is low- to mid-range.

Benefits

Using a random urine dipstick, and relying on the combination of leukocyte
esterase and nitrite to "rule out" the presence of UTI (i.e., both tests must be
negative) in combination with a lack of symptoms (i.e., the absence of symptoms reduces the pretest probability) reduces the need for further laboratory
investigations, and hence expedites clinical decision making, saves a second
clinic visit, and reduces the likelihood of unnecessary antibiotic use.

Challenges in practice

The approach to diagnosis of UTI is best achieved by using an algorithm that
gathers information on symptoms to advise on pre-test probability and then
using the urine test to produce a post-test probability—making the decision to
refer for laboratory analysis and/or give antibiotics at this point in the pathway.
Facilities for urinalysis and the availability of a competent POCT operator are
also important. The resources to perform POCT can be realized from the savings in the reduction of antibiotic use and of laboratory investigations.

Case Study 2: Chronic Disease Management—Self-Testing in Diabetes Mellitus

Background

Randomized controlled trials (RCTs) have shown the importance of tight
glycemic control in reducing the complications of diabetes and improving
both survival and quality of life (27,28). Thus self-monitoring of blood glucose (SMBG) has been widely adopted in the expectation that it will increase
the effectiveness of glycemic control strategies. Many studies have included
systematic reviews on the impact of SMBG, but they have provided only limited support for its use; this is reflected in the variations observed in published
guidelines (29–34).

The current American Diabetes Association (ADA) Guidelines for type 1
diabetes recommend blood glucose testing three times per day as part of an

integrated package of care with measurement of glycated hemoglobin (HbA1c) at least twice per year (32). For type 2 diabetes, the ADA Guidelines advise that SMBG may be useful in insulin-treated patients (31).

NICE also published guidelines for type 1 and 2 diabetes, in 2004 and 2002 respectively. For type 1 it recommended that SMBG should be part of integrated care and performed with a frequency that reflects glycemic targets, the type of insulin regimen used, and personal preference (32,33). For type 2 diabetes, NICE considered that SMBG did not appear to improve glycemic control as a stand-alone intervention and, while it may be appropriate to include SMBG in the package of care, the optimal frequency of testing had not been determined (34). A review by Coster et al. (30) pointed out the difficulty of obtaining robust evidence on the utility of SMBG (specifically) within the context of a complex disease management package, and the likely variation in the contribution of each element of the package in the individual studies. They recommended more randomized controlled trials as well as the use of data obtained from observational studies.

Clinical question

Does a care package that includes SMBG improve outcomes in patients with diabetes mellitus?

Setting

Patients with type 1 or type 2 diabetes using SMBG at home as part of a disease management care package to maintain tight glycemic control.

Evidence

The main evidence on the benefit of maintaining tight glycemic control comes from two studies, the Diabetes Control and Complications Trial (DCCT) (27) and the United Kingdom Prospective Diabetes Study (UKPDS) (28). The DCCT studied insulin-dependent diabetics and found that intensive blood glucose control reduced the risk of microvascular and neurological complications. The UKPDS studied patients with type 2 diabetes and showed that intensive blood glucose control with sulphonylureas or insulin substantially reduced the risk of microvascular but not macrovascular complications. Both studies demonstrated improvements in the HbA1c levels as a surrogate marker of glycemic control. The relationship between HbA1c and excess mortality risk has also been established (35).

The importance of a care package that is individualized for patients and that emphasizes education and personal empowerment has been underscored on many occasions; Norris et al., in a review of the evidence in a community setting, demonstrated the importance of self-management education (36).

The evidence for including SMBG in such management packages has been studied in a number of systematic reviews (30,37–39), with the first three

(30,37,38) focusing solely on data from randomized controlled trials. Looking primarily at the HbA1c level as an outcome measure, Coster et al. (30) found a pooled reduction of 0.576% (95%CI −1.073 to −0.061) for type 1 diabetics and concluded that SMBG should be part of a package of care. For type 2 diabetics the pooled reduction was only 0.25 (95%CI −0.61 to 0.10), and Coster et al. concluded that the effect was not sufficient to support SMBG. However the data were limited to a small number of studies with small populations, and the investigators recommended further studies that would include data from observational studies. Sarol et al. (37) studied type 2 non-insulin dependent diabetics and found a pooled decrease in HbA1c of 0.39% (95%CI −0.54 to −0.23) using a fixed effects model. Welschen et al. (38) undertook a similar systematic review and found an SMBG-related reduction in the HbA1c of 0.39% (95%CI −0.56% to −0.21%), with a similar degree of heterogeneity to that found by Sarol et al. Jansen (39) undertook a meta-analysis of the studies on non-insulin dependent type 2 diabetics using a Bayesian approach, including more studies, and found a pooled reduction in HbA1c of 0.40% (95%CI 0.07 to 0.70). A large observational study by Karter et al. (40) showed that SMBG was associated with a lowering of the HbA1c levels. Many papers have examined the role of SMBG; most of these papers appear to show some benefit. However it is difficult to draw strong conclusions because of the small patient numbers included in many of the studies and the variation in the care package used. In addition, many of the studies were conducted over relatively short time periods, and so the change in HbA1c should be viewed in that light. This naturally stresses the importance of continuing education and maintaining patient involvement in managing the disease.

Several papers have been written about the health economics of disease management in diabetes. While there is a cost associated with disease management protocols such as SMBG, the benefits of improved glycemic control appear in reduced acute hospital admissions and reduced incidents of secondary complications of diabetes (41–43).

Benefits

The benefits for patients are based on the improved quality of life and the reduction in risk associated with the complications of the disease, such as blindness, renal failure, and heart disease. Reduced complications will have a major economic benefit for healthcare providers and society as a whole.

Challenges in practice

The main challenges are associated with continuously providing a care package that ensures patient compliance with testing (including testing performance, e.g., quality control) and therapy protocols, as well as regular attendance at review meetings with a health professional. Many of these issues are beginning to be addressed through telemedicine initiatives, and the devolution of more care into the community setting (44).

Case Study 3: Chronic Disease Managem￼
Optimization for Anticonvulsant Drugs

Background

The expectation with any intervention is an impro
toms and a remission of the disease, with concurrent mi..
ated with any adverse reaction. Remission can be monitored using _
in some instances, diagnostic tests. In some cases, symptoms are of limit￼
due to lack of sensitivity and specificity, and in this situation a diagnostic test may
be of more value. The example of epilepsy and the use of anticonvulsant therapy
is a case in point, although there are many other examples (blood glucose in the
management of diabetes, prothrombin time testing in the management of warfarin
therapy, brain natriuretic peptide in the management of heart failure). Monitoring
anticonvulsant drug levels such as phenytoin is an essential part of the management of patients with epilepsy, as in many cases there is a narrow therapeutic circulating drug concentration window, and symptoms do not adequately reflect
optimal dosage. Persistent under- and over-dosage will result in adverse events.

Clinical question

Does the provision of antiepileptic drug measurement at the time of the clinic
visit improve the optimization of therapy?

Setting

An outpatient clinic managing patients with epilepsy.

Evidence

Patsalos et al. (45) undertook a small "before and after" study in a specialist
clinic with a control period of one year (n = 32 patients) and a test period of
one year (n = 53 patients). Measurements of phenytoin, carbamazepine, and
phenobarbitone were performed when patients arrived at the clinic. The results
were available after 15 minutes. Six patients met the criteria for inclusion in
the control and test periods, which specifically focussed on a newly implemented treatment program. Following the POCT intervention, there were
improvements in a number of the outcome measures:

- reduced time (weeks) from initial consultation to optimization of therapy
  [test period: mean, 4.5 (range, 2.0–9.0) vs. control period: mean, 22.9
  (range, 10.0–43.0), p < 0.05]
- fewer consultations required [test period: mean,1.8 (range, 1–2) vs. control
  period: mean, 4.0 (range, 3–5), p < 0.05]
- fewer drug dosage changes [test period: mean, 1.5 (range, 1–2) vs. control
  period: mean, 2.0 (range, 1–4)]
- fewer drug assays required [test period: mean, 2.5 (range, 1–4) vs. control
  period: mean, 4.8 (range, 3–7), p < 0.01].

...liot et al. (46) has shown that it can cost less to deliver a drug monitor-
...ervice in the clinic than in the laboratory service.

Benefits

Faster optimization of therapy benefits patient and clinician, providing an
objective assessment of compliance and offering greater convenience to both
parties. From an economic standpoint, while the drug measurements may be
more costly, these costs are readily offset by a reduction in clinic time and all
of the associated costs.

Challenges in practice

The main challenges to generating benefits from this use of POCT in this type
of scenario center around building the testing sequence into the clinic visit,
ensuring that the result is available at the time of the patient consultation, and
harvesting the resources resulting from the reduction in clinic visits to pay for
the additional testing costs.

Case Study 4: Chronic Disease Management—
Assessing Compliance in Diabetes Care

Background

The relationship between HbA1c and the mean blood glucose concentration
over a period of weeks is well known (47), and the role of HbA1c measurement in the management of diabetes is now well established in the literature
(48,49) as well as in clinical guidelines (32,33,50–53). Thus HbA1c is used as
a measure of glycemic control (54), as well as an indicator of the risk of developing the complications associated with poor glycemic control (55,56).

Clinical question

Does providing the HbA1c result during the consultation with the diabetes
management professional lead to improved health outcomes?

Setting

Hospital diabetes outpatient or primary care clinic.

Evidence

The majority of the many papers on POCT for HbA1c are concerned with the
technical performance of the devices. A recent paper compared all current POCT
devices for the quantitation of HbA1c in an outpatient clinic setting using nurse
operators (57). They found that the DCA 2000 (Bayer Diagnostics, Tarrytown,
NY, USA) gave the best precision in the hands of nurses who obtained a within-
day coefficient of variation (CV) of <2.5% and a between-day CV of <5%; this

data was comparable with laboratory staff operating the same system, and also with the performance of laboratory instrumentation. The authors concluded that the system was suitable for use in a clinic setting.

ECRI (formerly the Emergency Care Research Institute) recently reported on the performance of five HbA1c POCT systems. The evaluation focused on analytical performance and ease of use of several POCT devices and provided purchasing guidance for different types of healthcare facilities (58).

Reports on five trials, including a primary research health technology assessment, are available on POCT for HbA1c. Cagliero et al. (59) reported on an RCT involving 201 type 1 and type 2 diabetic patients attending a secondary care diabetes center. The patients were randomized to either a consultation in which the clinician received immediate feedback on the HbA1c result, or to the routine service when the result came back from the laboratory at a later date. The HbA1c results decreased in the POCT group at 6- and 12-month follow up (-0.57 ± 1.44 and $-0.40 \pm 1.65\%$ respectively; $p < 0.01$) with no significant change in the control group (-0.11 ± 0.70 and $-0.19 \pm 1.16\%$ respectively). In a controlled trial, Thaler et al. (60) studied 1138 diabetic individuals attending an urban diabetes center and found that more appropriate management was achieved in patients whose HbA1c results were generated by POCT ($p < 0.0001$), with fewer changes in treatment when the HbA1c was $<7.0\%$, and more when it was $>7.0\%$. Over the follow-up period (a maximum of 7 months) the HbA1c levels rose more in the conventional testing group compared to the POCT group. Miller et al. (61) performed a similar study of 597 patients with diabetes in an urban primary care setting. They found that treatment intensification was greater in those receiving POCT, and in those who had higher baseline HbA1c levels. Furthermore, the HbA1c levels decreased over the course of the study in the POCT group (8.4% to 8.1%, $p = 0.04$, compared with 8.1% to 8.0%, $p = 0.31$). Kennedy et al. (62) reported on the impact of POCT for HbA1c, together with algorithmic titration of basal insulin, on glycemic control in patients with type 2 diabetes. Active titration yielded greater reductions in HbA1c, and POCT for HbA1c was associated with an increase in the proportion of patients achieving an HbA1c $<7.0\%$ (41% vs. 36% in the control arm, $p < 0.0001$).

In a retrospective study of two cohorts of patients attending clinics—one using POCT for HbA1c and the other receiving results from a laboratory at a later date—Grieve et al. (63) found the mean HbA1c was significantly higher in the cohort receiving the results at a later date (8.66 ± 0.056 vs. 7.79 ± 0.058; $p < 0.001$). Grieve et al. also investigated the feasibility of introducing POCT for HbA1c together with other biochemical tests in 599 individual patient clinic visits. They found that more clinical management changes were made in the group of patients treated with POCT HbA1c than those treated with conventional testing, where the results were available at some later date (23 vs. 18%). They also studied patient and clinician satisfaction using questionnaires and found increased satisfaction levels in those clinic visits that incorporated POCT. Ferenczi et al. (64), in a retrospective review of medical records of new referrals, found that patients receiving care using immediate HbA1c results showed a greater decrease in HbA1c compared to those who received the result two days later (1.03 ± 0.33% vs. 0.33 ± 0.83%). Holman et al. (65) conducted a "before and after study" that involved sending patients small collection tubes, to be returned with their blood specimen. Seventy-four percent of the collection tubes sent out over one year were usable upon return, and this was associated with a reduction of the mean HbA1c result of 0.8% compared with the previous year (p < 0.001). As pointed out in an earlier chapter, it is also possible to bring the patient to the clinic for phlebotomy the week before a consultation, although this may be less convenient for the patient.

The role of education in diabetes management, notably that related to blood glucose measurement, diet, and lifestyle, is crucial. Patients' awareness of their HbA1c result is also important. Heisler et al. (66) performed a cross-sectional survey of type 2 diabetics who had HbA1c levels checked in the previous six months. A significant number of the cohort (66%) did not know their HbA1c result and only 25% reported it correctly. Participants who knew their most recent HbA1c result were more likely to accurately assess their diabetes control (adjusted odds ratio 1.59, 95% CI 1.05–2.42) and to better understand their diabetes care (p < 0.001). Levetan et al. (67) undertook a randomized trial to investigate the impact of personal knowledge of the HbA1c result and associated goals on the ensuing HbA1c results. In the 128 patients who completed the study, there was an absolute reduction of 1.69% in the HbA1c result in the intervention arm of the study in patients with a baseline HbA1c result ≥ 7.0%, compared with the control arm where there was a reduction of 0.77% (p = 0.032).

Economic assessments of the use of POCT for HbA1c are limited. The study by Cagliero et al. measured the utilization of a wide range of healthcare resources, including outpatient visits and contact time with staff, and found that POCT did not significantly change these measures (59). Grieve et al. (63) found the costs of POCT for HbA1c higher than those for the laboratory-provided service; when a laboratory analyzer was taken down to the clinic and run by a laboratory technologist, the costs were marginally higher than the conventional laboratory service. However from an analysis of the retrospective cohort study, they found a reduction in clinic visits using the POCT modality (from 2.28 visits per year per patient to 1.81), which helped ameliorate the increased cost of testing.

Benefits

Using POCT to provide HbA1c results during the clinic visit appears to improve HbA1c levels, presumably as a consequence of the result becoming part of the consultation. The longer-term impact of POCT strategies must be extrapolated from other studies that have demonstrated the benefits of maintaining improved glycemic control and of reducing risk associated with lower HbA1c levels.

Challenges in practice

The two main difficulties in applying POCT are associated with building the testing component into the logistics of the clinic care pathway and reallocating the resources to account for increased test costs and reduced clinic support costs. If the clinic is held in a primary care setting, then there is also the challenge of shifting resources from the secondary to the primary care provider. However, this should be embraced in the dialogue around payment by results (or payment for performance) and practice-based commissioning, strategies that several countries have adopted for controlling costs and improving outcomes.

Case Study 5: Chronic Disease Management— Assessing Compliance in Anticoagulation Care (INR Testing)

Background

In the past, most patients on coumadin (warfarin) therapy had to attend an outpatient clinic in a secondary care setting. However they are now able to visit their primary care center or even perform the tests in their own home or workplace, because a number of POCT devices for prothrombin time (PT) that assess the integrity of the extrinsic and common pathways of coagulation are available.

The most common causes of an abnormal PT are a change in the efficacy of oral anticoagulation therapy, vitamin K deficiency, liver dysfunction, and disseminated intravascular coagulation. In addition, there is a significant level of interpersonal variation in response to coumadin (warfarin). Consequently regularly checking PT status is important.

One of the main issues that have concerned physicians regarding POCT systems for PT is the variation in the quality and sensitivity of reagents. The use of the International Normalized Ratio (INR), which is calculated from the PT reading and the international sensitivity index, can be used to nullify the effect of the majority of differences in reagent sensitivities. However, it is still important to ensure that POCT devices for INR testing give analytically valid results.

Clinical question

Does assessment of PT status using INR testing lead to improve compliance with coumadin (warfarin) therapy when testing is performed either in primary care or by the patient?

Setting

Predominantly in primary care, and increasingly the workplace or the patient's home. There are some specific hospital settings.

Evidence

Several studies have demonstrated comparable performance between POCT and laboratory assessment of PT status as a result of the use of INR testing (68,69). Patient questionnaires have also shown greater satisfaction with the community-based POCT monitoring program.

A structured educational program is considered vital to any self-management program for patients receiving oral anticoagulation. Sawicki et al. (70), in a randomized single-blind multicenter trial involving 179 patients, showed that deviation from the mean of the INR target range was significantly lower in the intervention group (trained) at 3 months (squared INR deviation 0.59 vs. 0.95; $p < 0.001$) and at 6 months (0.65 vs. 0.83; $p = 0.03$) compared with the control group. The intervention group also had INRs within the target range more often, with less frequent suboptimal INRs (32% vs. 50% at 3 months and 34% vs. 48% at 6 months) compared with the control group. The treatment satisfaction scores were also found to be significantly higher in the intervention group.

A number of "test-and-treat" algorithms have been described. These algorithms are typically linked with some form of computerized dosage protocol. Thus Fitzmaurice et al. evaluated a computerized decision support tool in one primary care practice, while in a second practice patients were randomized to receive either computerized decision support or conventional care using the local laboratory (71). INR control significantly improved (from 23% to 86%, $p < 0.001$) at the first center, which used computerized decision support. At the second center, where patients were randomized to receive decision support or conventional care, logistic regression showed improved INR control ($p < 0.001$) in patients receiving decision support but not in patients managed by conventional care. In the decision support patients, there was also a notable extension in the length of time before patients were recalled to the practice, with no difference in the number of adverse events. In a larger randomized controlled trial involving more health centers (nine intervention and three control), Fitzmaurice et al. (72) showed that the intervention group spent more time in the target INR range ($p = 0.008$). The authors concluded that this approach could be used in primary healthcare centers in developed healthcare systems.

Poller et al. (73) performed a randomized study of computerized support involving 285 patients in five European centers. The data from all centers were combined, and the investigators found that more patients using computerized support achieved the target INR ($p = 0.004$), and that their mean time within range was greater (63.3% vs. 53.2%, SD 28.0 and 27.7 respectively) compared with the conventional treatment group. Manotti et al. undertook a multicenter randomized controlled trial of computer-aided management involving 1251 patients, of which 335 had only been treated for three months (74). Patients were randomized to receive either computerized support, or dosing based on the usual physician support. The investigators found faster stabilization of therapy ($p < 0.01$) in the computerized support group, and this group spent more time within the therapeutic range (p < 0.001). As in other studies, the improvement was mainly due to less time spent with a sub-therapeutic level of oral anticoagulant. There was also a significant reduction in the number of appointments required (p < 0.001). Kovacs et al. assessed a coumadin initiation nomogram in 105 consecutive referrals to outpatients (75). In their study, 83% of the patients had reached the treatment goal in five days and 98% had reached the goal in eight days. Furthermore there were no reported incidents of bleeding, and in only six cases did the INRs exceed 4.5. The authors concluded that the program was worthy of further study.

Siebenhofer et al. undertook a systematic review of studies of self-management of oral anticoagulation and found nine relevant randomized controlled trials, but only four met the inclusion criteria (76). Comparing the four studies, the investigators found no difference in oral anticoagulation control between self-management care and management by a specialized anticoagulation clinic. In comparison with routine care by general practitioners, the authors found that self-management care to be better. Data from two studies showed that self-management could clearly improve treatment-related quality of life. The authors concluded that self-management oral anticoagulation treatment is safe and improves treatment-related quality of life. However, the authors did note that there were no valid long-term studies that had demonstrated reduction of bleeding and thromboembolic events, the major risks associated with anticoagulation therapy.

More recently, Heneghan et al. (77) reported a systematic review and meta-analysis of self-monitoring of oral anticoagulation, finding 14 randomized trials that met their inclusion criteria. The pooled estimates showed significant reductions in thromboembolic events (odds ratio 0.45, 95% CI 0.30–0.68), all-cause mortality (0.61, 0.38–0.98), and major hemorrhage (0.65, 0.42–0.99). They also found that trials of combined self-management (monitoring and therapy adjustment) showed significant reductions in thromboembolic events (0.27, 0.12–0.59) and death (0.37, 0.16–0.85), but not major hemorrhage (0.93, 0.42–2.05). A total of 11 trials reported improvements in the mean proportion of international normalization ratios in range. The authors concluded that self-management improves the quality of oral anticoagulation, and that patients who monitor and adjust therapy have fewer thromboembolic events and lower mortality than those who self-monitor alone. They noted, however, that self-monitoring is not feasible for all patients; it requires recognition and education of suitable candidates—a conclusion drawn in a number of studies.

Several studies have focused on patients with specific anticoagulation needs, for example, patients receiving heart valve replacements. Sidhu and O'Kane undertook a two-year prospective randomized trial of self-managed anticoagulation therapy in 100 patients following valve replacement (78). The self-managed group demonstrated better control as judged by number of test results in range (67.6% vs. 58.0%) and time-in-range (76.5% vs. 63.8%), compared with the conventionally managed group. There was no difference in morbidity or mortality between the two groups. Kortke et al. studied self-management in a group of patients given a mechanical heart valve replacement (79). A randomization scheme was used to assign patients to self-management or to the control group managed by a family practitioner. A higher incidence of INR values within range occurred in the self-managed group compared to the control group (80% vs. 62%) and the incidence of complications (hemorrhages and thromboembolic events) in the self-managed group was also reduced (p < 0.05). Christensen et al. in another small study followed 24 patients with mechanical heart valve replacements for a total of four years (80). They found that the self-managed patients were within a target INR range for a mean of 78% (range 36.1%–93.9%) of the time, compared with 61.0% (range 37.4%–2.9%) in the control group. The investigators concluded that self-management was a feasible and safe approach to anticoagulation therapy. In a later study Koertke et al. reported data on 1,818 patients who had received a mechanical heart valve replacement (81). The authors performed a randomized controlled trial, assigning patients to specific treatment target INRs depending on the valve replaced. The investigators concluded that the early initiation of INR self-management after mechanical heart valve replacement enabled patients to stay within a lower and smaller INR target range. In addition, the reduced anticoagulation level that could be achieved resulted in fewer grade III bleeding complications without increasing thromboembolic event rates.

Benefits

The benefits of POCT for PT status have only become possible as a result of technological advances that ensure comparability between POCT and laboratory performance, a requirement for all POCT but uniquely problematic in the case of PT testing and the use of the INR.

Clinical benefits of POCT for PT status include more patients achieving and staying within the target INR range for longer periods of time; fewer complications (hemorrhage or thromboembolic events); fewer dosage changes; and fewer clinic visits. In addition, patient and clinician satisfaction increases. There is also some evidence that it is less costly to deliver a POCT-based anticoagulation service than a hospital-based service, both from the viewpoint of the patient as well as of the provider. The economic benefits to be achieved from the reduction in hospital clinic facilities may allow resources to be redirected to increased investment in POCT.

Challenges in practice

There are certain important features to achieving a good INR testing service. It is important to ensure consistency of performance of the POCT system with respect to INR results. This requires careful procurement of reagents and good quality management procedures, including regular quality control. In addition, training of operators and patients when using self-management is vital, as is recognizing that not all patients may be able to cope with a POCT system.

Case Study 6: Emergency Department–Improving Patient Triage

Background

Many of the early developments in POCT in the emergency department (ED) were based on the supposition by ED physicians that providing results faster enabled clinical decisions to be made more quickly and interventions to be initiated earlier. It may have simply been a case of allowing more effective utilization of their time (82). However, several studies, including a randomized trial, showed that POCT did not significantly decrease turnaround times (TATs) or length of stay (LOS) in the ED (83–85). Many of these early studies were based on using POCT systems to measure blood electrolytes and gases; these may not be the tests that ED physicians need to answer questions related to helping with early diagnosis and triage. Thus this example underscores the importance of knowing what decision the physician needs to make and then, whether rapidly providing the relevant tests results enables earlier decision making and clinical intervention.

Clinical question

Will the availability of POCT in the ED lead to more rapid triage of patients through the ED, and to earlier interventions?

Setting

Patients presenting to the ED.

Evidence

Murray et al. (85) performed one of the first studies to show any improvement in triage of patients through the ED with a reduced length of stay (LOS). The investigators found a reduction in the median LOS from 4.22 hours (interquartile range 3.04–5.47) using the laboratory service to 3.28 hours (interquartile range 2.28–5.30) using POCT, for electrolytes and blood gases. However, they noted that many patients had to wait for results from laboratory-based investigations before a clinical decision could be made. The reduction in LOS was only achieved in patients whose normal POCT test results led to an earlier discharge.

Lee-Lewandrowski et al. (86) approached the issue from a different perspective because of pressure on the ED, which led to patients being referred to other hospitals. They investigated the use of tests that the ED physicians considered might help reduce LOS as a result of shorter turnaround times (TAT) and consequent quicker decision making. They showed that POCT reduced the TAT for results for blood glucose, cardiac markers, urine pregnancy testing, and simple urinalysis (an average of 83%), which enabled a reduction in the

ED LOS (mean 41.3 min, p < 0.006) when the results could be combined together to make clinical decisions sooner. In addition to improved TAT and LOS, ED physician satisfaction significantly improved (p < 0.001). This was attributed to setting up a POCT kiosk in the ED, although the productivity of the staff was low by comparison with the laboratory, and the cost per test was $2.94 in the laboratory and $19.20 in the kiosk. Clearly, adding additional tests could improve productivity as well as increase the potential utility by applying POCT to more of the admissions. Even at the average saving in LOS of 41.3 min, the total saving for the workload over a year (25,812 tests) amounted to the equivalent of 2.0 ED beds.

One of the lessons of this and other studies in this area is the difficulty of assessing the economic impact of POCT due to the complex nature of ED work and its interrelations with other parts of the organization. However, POCT in this setting could likely reduce the bed requirement with a trade-off between saving 41.3 min per patient and an incremental testing cost of $16.26. Note also that implementing the POCT kiosk and other related organizational changes led to a 27% reduction in the number of patients who had to be diverted to other hospitals, thereby increasing income to the organization.

Several studies have examined the utility of POCT for cardiac markers in the ED. Ng et al. (87) found that all patients with chest pain and suspected myocardial infarction could be effectively diagnosed within 90 minutes of admission (positive predictive value 47%, negative predictive value 100%) and coronary care unit admissions were reduced by 40%. A total of 90% of patients with negative cardiac markers and ECG findings at 90 minutes were discharged, and only one returned with an infarction (0.2%) within the next 30 days. McCord et al. reported similar results from a similar testing strategy, with sensitivity and negative predictive values of 96.9% and 99.6% respectively (88). Zarich et al. also showed that using a rapid cardiac marker service improved utilization of hospital resources and reduced costs (89). Ohman et al. showed that a POCT device for cardiac markers could be used effectively for risk stratification of patients, thus enabling more informed patient triage through a chest pain evaluation unit (90).

Cardiovascular risk may also result from deep vein thrombosis. Kline et al. (91), in a prospective non-interventional study, showed that a simple whole blood agglutination test for D-dimer, together with an assessment of alveolar dead space, could exclude pulmonary embolism in patients admitted to the emergency room (sensitivity 98.4%, 95% confidence limits 91.6–100%; specificity 51.6%, 95% confidence limits 46.1–57.1%). The authors concluded that a normal D-dimer test result was associated with a low prevalence of pulmonary embolism. Bates et al. (92) in a similar study concluded that a simple D-dimer test was useful for ruling out pulmonary embolism in patients with suspected venous thromboembolism. Brown et al. (93) in a meta-analysis found that the simple turbidimetric assays for D-dimer were sensitive but not specific for the detection of pulmonary embolism, with results very much in line with those of Kline (91) and Bates (92). Kline et al. then investigated the use of a rapid "rule out" protocol (based on their earlier work) (94) and found that it doubled the rate of screening in the ED, had a false negative rate of <1%, did not increase the rate of pulmonary vascular imaging, and reduced the length of stay. Roy et al. (95) in a recent systematic review of strategies for the diagnosis of pulmonary embolism pointed out the considerable variation in the accuracy of diagnostic strategies, indicating that the "rule out" approaches in patients with a low pre-test probability were associated with a post-test probability of <5%.

The point has already been made that the benefits from POCT only become clear when clinical practice changes to make use of the rapid response. This point is borne out by Nichols et al. (3), who prospectively studied 216 patients requiring diagnostic laboratory testing for coagulation (prothrombin time/activated partial thromboplastin time) and/or renal function (urea, creatinine, sodium, and potassium) before elective invasive cardiac and radiological procedures. They studied workflow using current practice at the time, implementation of POCT, therapeutic decisions based on POCT results, and optimization of workflow around the availability of POCT. Using current practice (central laboratory results), 44 percent of results were not available before the scheduled time for procedure. In patients needing renal function tests, POCT decreased patient waiting times, but in patients needing coagulation tests, the waiting times only improved when systematic changes were made in workflow. The authors concluded that the benefits of POCT only become evident when systematic changes in the care pathway were implemented.

Benefits

"Appropriate" POCT in the ED is testing that answers the clinical questions posed by individual patients' circumstances. When that is the case, POCT can facilitate more rapid clinical diagnosis and decision making, as well as more efficient triage of patients.

Challenges in practice

The wide spectrum of patients presenting to the ED means that a wide range of tests is likely to be required. Thus while the productivity of testing in the ED may not be high, it will be offset by the savings in patient stay, and possibly reduce the need to refer patients to other hospitals. The two key challenges identified from the studies of POCT in the ED are the changes in clinical practice that are required in order for clinicians to be able to act on POCT results, and the transfer of resources from other secondary care facilities where utilization is reduced—for example, coronary care unit admissions and bed stay—into POCT facilities.

Case Study 7: Operating Room—Improving the Quality of Intervention

Background

The assessment of blood gas, electrolyte, and vital sign status is now an accepted part of intra-operative monitoring, but there are other instances where monitoring of biochemical and hematological parameters has clinical and economic benefits. It is now well established, for example, that ionized calcium should be monitored during liver transplantation in order to assess the effect of the citrate burden from transfused blood, thereby reducing the risk of a cardiac arrest (96). More recently there has been a great deal of interest in the monitoring of coagulation status, particularly during cardiopulmonary bypass procedures. Thus several studies have shown the opportunity for more rapid decision making with reduction in post-operative complications, reduced blood loss (and therefore utilization of transfused blood products), and reduced need for high dependency post-operative care (97,98).

Rapid test results can also be used to assess the effectiveness of surgery, as in the case of intra-operative parathyroid hormone (PTH) monitoring during parathyroidectomy.

Clinical question

Will POCT for the intra-operative measurement of parathyroid hormone improve the outcomes from parathyroidectomy?

Setting

Operating room.

Evidence

Irvin et al. have shown that measuring the circulating PTH level during parathyroidectomy can reduce the need for re-operation by ensuring that all of the tissue is removed at the initial operation (99). Their review of the literature showed that despite good preoperative localization of the gland, the parathyroidectomy failure rate was between 5% and 10% and possibly higher in less experienced hands. They compared the outcomes in 50 consecutive patients undergoing more difficult secondary parathyroidectomy with and without the support of intra-operative PTH measurements, using return to normocalcemia as an outcome measure. In patients who were operated on with the aid of intra-operative PTH measurement, calcium levels returned to normal in 31 of 33 patients. Together with good preoperative localization studies, 17 patients underwent successful straightforward parathyroidectomies as predicted by their PTH levels. In the other 14 patients, the PTH assay helped in the localization of the gland with differential venous sampling, a technique in which the increase in hormone secretion is measured after massage of specific areas. In the patients who underwent surgery without the aid of a PTH assay, 4 of 17 patients (24%) remained hypercalcemic after extensive re-exploration. Thus with the intraoperative hormone assay used to facilitate localization and confirm excision of all hyperfunctioning tissue, the success rate of parathyroidectomy improved from 76% to 94%. Their experience was borne out in a much larger series of 421 patients, including those with multiglandular disease (100).

Subsequently, Chen et al. (101) showed that using the PTH assay together with localization of the gland using sestabmibi-SPECT and cervical block anesthesia transformed parathyroidectomy into a day case procedure. The study compared the use of the aforementioned procedure on 33 patients with a single adenoma to the use of bilateral exploration and general anesthesia on 186 patients. The day case procedure benefited the patient and the hospital budget.

Benefits

Clearly POCT improved the success rate of the surgical procedure, increased patient and clinician satisfaction, and reduced the risks associated with general anaesthesia and hospital stay. In the study by Chen et al., the hospital stay was reduced from 1.8 ± 0.1 to 0.3 ± 0.2 days, $p < 0.001$, and the cost from 6328 ± 292 to 3174 ± 386 (1999 figures), $p < 0.001$.

Challenges in practice

Apart from the necessary training in the use of the new procedures, the key challenge is ensuring that the resources are available to provide the diagnostic techniques during surgery.

Case Study 8: Intensive Care Unit—Improving Patient Management and Outcomes

Background

Biochemical and hematological testing are well established in the management of patients in the intensive care unit (ICU). In particular, blood gases and electrolytes are probably as important as the vital signs in many patients; several of these measurements are used to predict weaning from mechanical ventilation in intensive care patients (102). However, an increasing number of tests can now be used to answer specific questions associated with risk as well as disease status. Halpern et al. (103) and Kilgore et al. (104) used modeling studies to conclude that in addition to the clinical benefits of a rapid delivery of results, there were considerable economic benefits to the use of POCT. Kilgore et al. in fact suggested that annual savings in excess of $250,000 (at 1996 prices) could be made in a large hospital, despite the technology costs being greater.

The intensive care unit was one of the first areas of the hospital to have access to POCT as blood gas, and then electrolyte systems, became available for use. Prior testing had been subject to the vagaries of the hospital transport system and the use of satellite laboratories. Furthermore the knowledge of blood gas exchange, and the mechanism for maintaining hydrogen ion and electrolyte status, together with the poorer morbidity and mortality associated with changes in the circulating levels of these ions, established a strong "scientific" case for their routine (and "stat") use. As a consequence there is little evidence-based outcomes data on the use of POCT for these common analytes, and it would be unethical to perform such a study when their use is embedded in the modern standard of care. However data on additional analytes is beginning to appear.

Clinical question A

Will blood lactate measurement improve outcomes in ICU patients?

Setting

Intensive care unit in a pediatric hospital.

Evidence

In a study of goal-directed medical therapy after congenital heart surgery, Rossi et al. (105) studied the use of blood lactate measurements taken serially for 24 hours after heart surgery in a case control study involving more than 2,000 patients. Therapy was adjusted according to the blood lactate values and trends. The mortality was lower in the group of patients (n = 710) in whom the blood lactates were measured (1.8 vs. 3.7%, p = 0.02), with the greatest reduction seen in neonates (3.4% vs. 12.0%, p = 0.02), with no difference in older children. The mortality was also lower in patients undergoing higher-risk operations (Risk Adjustment for Congenital Heart Surgery-1 [RACHS-1] categories 3–6) (3.0% vs. 9.0%, p = 0.006), with no difference in those patients undergoing lower risk procedures.

Benefits

There are demonstrable improved clinical outcomes from close monitoring of blood lactate, and obvious benefits in patient and clinician satisfaction. It is likely that there will also be benefits in terms of overall length of stay with improved recovery periods.

Challenges in practice

The use of POCT in the ICU is well established, and so the only challenge is likely to be in making the additional resources available for purchase of the POCT blood lactate measurements.

Clinical question B

Will close glycemic control improve outcome in patients on the ICU?

Setting

Intensive care unit.

Evidence

There is now a growing literature on the merits of tight glycemic control in patients in the ICU (106,107). The evidence suggests that close control of the blood glucose level rather than infused insulin dosage is more closely associated with beneficial effects (108). The benefits include improved mortality, reduced infection rates, and reduced length of stay in the intensive care unit (109). A number of studies have employed treatment algorithms to maintain the blood glucose levels between strict limits, all showing an improvement in a number of metrics including reductions in infection rates, utilization of blood products, mortality, and length of stay (110–112).

Kanji et al. (113) sounded a note of caution in a study of three different approaches to POCT for blood glucose in the intensive care unit. They showed significant differences between the use of a blood glucose meter and the reference laboratory method; better agreement was shown with the use of a blood gas/chemistry benchtop analyzer system. The differences led to clinical disagreement on the insulin dosage that should be used.

Benefits

The evidence would suggest that there are significant clinical and economic benefits from the use of tight glycemic control in the ICU patient. This has therefore been proposed as the standard of care for patients in the ICU.

Challenges

The wide acceptance of blood parameter monitoring in the ICU means that the implementation of protocols for the tight control of blood glucose is gaining wide acceptance. The existence of treatment algorithms and protocols means that the risk of hypoglycemia is likely to be minimized.

Clinical question C

Will measurement of the urine albumin-creatinine ratio predict outcome in patients admitted to the ICU?

Setting

Intensive care unit in a large teaching hospital.

Evidence

Allison et al. (114) showed that microalbuminuria in the ICU patient can be used as an indicator of increased vascular permeability. Gosling et al. (115) have since shown in 431 consecutive admissions to the ICU that the urine albumin changes rapidly during the first six hours following ICU admission and predicts ICU mortality and inotrope requirement as well or better than APACHE II and SOFA scores. Further work is now in progress to assess the clinical utility of these observations.

Benefits

The simple measurement of the urine albumin-creatinine ratio could be a valuable tool for monitoring progress in the ICU, and for assessing prognosis; additionally this may produce benefits similar to those observed with tight glycemic control.

Challenges

At this point the challenge is to find the clinical decisions that would benefit from the use of this simple measurement.

SUMMARY AND RECOMMENDATIONS

A wide range of reliable technology is now available for POCT—providing that proper organization and management of the service are maintained. The provision of rapid test results through the use of POCT enables faster clinical decision making and earlier interventions. These are often gained through a change in clinical practice and the way that care is delivered. An additional benefit is that a more patient-centered approach to healthcare is achieved by providing care closer to the patient.

The cited case studies have illustrated the opportunity to improve the clinical outcomes with gains in improved morbidity and mortality. This is true for a variety of different primary and secondary care settings.

What can we learn from these POCT studies?

- Patient self-testing can improve outcomes.
- Education and training are important for all users of POCT.
- Knowledge of results can help patients manage their disease.
- POCT can improve clinical outcomes in a number of settings.
- POCT can improve economic outcomes in a number of settings.
- Though POCT may be more expensive, the economic gains outweigh the additional investment.
- A change in the way care is delivered is invariably required to gain all of the benefits from POCT.

REFERENCES

1. Guyatt G, Haynes B, Jaeschke R, et al. Introduction: the philosophy of evidence-based medicine. In: Guyatt G, Rennie D, eds. User's guide to the medical literature. A manual for evidence-based clinical practice. Chicago: JAMA and Archive Journals, American Medical Association, 2002:3–12.
2. Lundberg GD. The need for an outcomes research agenda for clinical laboratory testing. JAMA 1998;280:565–6.
3. Nichols JH, Kickler TS, Dyer KL, et al. Clinical outcomes of point-of-care testing in the interventional radiology and invasive cardiology setting. Clin Chem 2000;46:543–50.
4. Price CP, St John A, Hicks JM, eds. Point-of-care testing, 2nd ed. Washington, DC: AACC Press, 2004:488pp.
5. The National Academy of Clinical Biochemistry. Laboratory medicine practice guidelines: draft guidelines on evidence-based practice for POCT. Available at http://www.nacb.org/lmpg/poct_lmpg_draft.stm (Accessed February 12, 2006).
6. Foxman B, Barlow R, D'Arcy H, et al. Urinary tract infection: self-reported incidence and associated costs. Ann Epidemiol 2000;10:509–15.
7. Foxman B. Epidemiology of urinary tract infections: incidence, morbidity, and economic costs. Am J Med 2002;113 Suppl 1A:5S–13S.
8. Emmerson AM, Enstone JE, Griffin M, et al. The second national prevalence survey of infection in hospitals—overview of the results. J Hosp Infect 1996;32:175–90.
9. Neild GH. Urinary tract infection. Medicine 2003;21:85–90.
10. Christiaens TC, De Meyere M, Verschraegen G, et al. Randomised controlled trial of nitrofurantoin versus placebo in the treatment of uncomplicated urinary tract infection in adult women. Br J Gen Pract 2002;52:729–34.
11. Fahey T, Webb E, Montgomery AA, et al. Clinical management of urinary tract infection in women: a prospective cohort study. Fam Pract 2003;20:1–6.
12. SMAC. Standing Medical Advisory Committee (SMAC) report: the path of least resistance. Occasional Report 1998. Public Health Laboratory Service, London, UK. Available at http://www.advisorybodies.doh.gov.uk/smac1.htm (Accessed February12, 2006).
13. Fenwick EA, Briggs AH, Hawke CI. Management of urinary tract infection in general practice: a cost-effectiveness analysis. Br J Gen Pract 2000;50:635–9.
14. Colodner R, Rock W, Chazan B, et al. Risk factors for the development of extended-spectrum beta-lactamase-producing bacteria in nonhospitalized patients. Eur J Clin Microbiol Infect Dis 2004;23:163–7.
15. Woodford N, Ward ME, Kaufmann ME, et al. Community and hospital spread of *Escherichia coli* producing CTX-M extended-spectrum beta-lactamases in the UK. J Antimicrob Chemother 2004;54:735–43.
16. Smith HS, Hughes JP, Hooton TM, et al. Antecedent antimicrobial use increases the risk of uncomplicated cystitis in young women. Clin Infect Dis 1997;25:63–8.
17. Graham JC, Galloway A. ACP Best Practice No. 167: the laboratory diagnosis of urinary tract infection. J Clin Pathol 2001;54:911–9.
18. Smith PJ, Morris AJ, Reller BL. Predicting urine culture results by dipstick testing and phase contrast microscopy. Pathology 2003;35:161–5.
19. Etherington IJ, James DK. Reagent strip testing of antenatal urine specimens for infection. Br J Obstet Gynaec 1993;100:806–8.

20. Rouse DJ, Andrews WW, Goldenberg RL, et al. Screening and treatment of asymptomatic bacteriuria of pregnancy to prevent pyelonephritis: a cost-effectiveness and cost-benefit analysis. Obstet Gynecol 1995;86:119–23.

21. National Collaborating Centre for Women's and Children's Health. Antenatal care: Routine care for the healthy pregnant woman. Clinical Guideline (NICE) 2003;81. Available at http://www.rcog.org.uk/index.asp?PageID=693 (Accessed February 12, 2006).

22. Gorelick MH, Shaw KN. Screening tests for urinary tract infection in children: a meta-analysis. Pediatrics 1999;104:e54.

23. Huicho L, Campos-Sanchez M, Alamo C. Meta-analysis of urine screening tests for determining the risk of urinary tract infection in children. Pediatr Infect Dis J 2002;21:1–11.

24. Hurlbut TA III, Littenberg B. The diagnostic accuracy of rapid dipstick tests to predict urinary tract infection. Amer J Clin Pathol 1991;96:582–8.

25. Deville WL, Yzermans JC, van Duijn NP, et al. The urine dipstick test useful to rule out infections. A meta-analysis of the accuracy. BMC Urol 2004;4:4.

26. St John A, Boyd JC, Lowes AJ, Price CP. The use of urinary dipstick tests to exclude urinary tract infection: a systematic review of the literature. Am J Clin Path 2006 (accepted for publication).

27. The Diabetes Control and Complications Trial Research Group. The effect of intensive treatment of diabetes on the development and progression of long-term complications in insulin-dependent diabetes mellitus. New Engl J Med 1993; 329:977–86.

28. UK Prospective Diabetes Study (UKPDS) Group. Intensive blood-glucose control with sulphonylureas or insulin compared with conventional treatment and risk of complications in patients with type 2 diabetes (UKPDS 33). Lancet 1998;352:837–53.

29. Faas A, Schellevis EG, Van Eijk JTM. The efficacy of self-monitoring of blood glucose in NIDDM subjects: a criteria-based literature review. Diabetes Care 1997;20:1482–6.

30. Coster S, Gulliford MC, Seed PT, et al. Monitoring blood glucose control in diabetes mellitus: a systematic review. Health Technology Assessment 2000;4:1–93.

31. American Diabetes Association. Standards of medical care in diabetes. Diabetes Care 2004;27:15S–35S.

32. National Institute for Clinical Excellence. Type 1 diabetes: diagnosis and management of type 1 diabetes in children, young people and adults. Available at http://www.nice.org.uk?CGO15NICEguideline (Accessed February 12, 2006).

33. National Institute for Clinical Excellence. Diabetes - type 2 (update). Available at http://www.nice.org.uk/page.aspx?o=264875 (Accessed February 12, 2006).

34. Alberti KGMM, Gries FA, Jervell J. European Diabetes Policy Group. A desktop guide to type 2 diabetes mellitus. Diabet Med 1994;1:899–909.

35. Khaw KT, Wareham N, Luben R, et al. Glycated haemoglobin, diabetes, and mortality in men in Norfolk cohort of European prospective investigation of cancer and nutrition (EPIC-Norfolk). BMJ 2001;322:15–8.

36. Norris SL, Nichols PJ, Caspersen CJ, et al. The effectiveness of disease and case management for people with diabetes. A systematic review. Am J Prev Med 2002;22(4 Suppl):15–38.

37. Sarol JN Jr, Nicodemus NA Jr, Tan KM, Grava MB. Self-monitoring of blood glucose as part of a multicomponent therapy among non-insulin requiring type 2 diabetes patients: a meta-analysis (1966–2004). Curr Med Res Opin 2005;21: 173–84.

38. Welschen LM, Bloemendal E, Nijpels G, et al. Self-monitoring of blood glucose in patients with type 2 diabetes who are not using insulin: a systematic review. Diabetes Care 2005;28:1510–7.

39. Jansen JP. Self-monitoring of glucose in diabetes mellitus: a Bayesian meta-analysis of direct and indirect comparisons. Curr Med Res Opin 2006;22:671–81.

40. Karter AJ, Ackerson LM, Darbinian JA, et al. Self-monitoring of blood glucose levels and glycemic control: the Northern California Kaiser Permanente Diabetes Registry. Am J Med 2001;111:1–9.

41. Diabetes Control and Complications Trial Research Group. Lifetime benefits and costs of intensive therapy as practiced in the Diabetes Control and Complications Trial. JAMA 1996; 276: 1409–15.

42. Gray A, Raikou R, McGuire A, et al. Cost effectiveness of an intensive blood glucose control policy in patients with type 2 diabetes: economic analysis alongside randomised controlled trail (UKPDS 41). BMJ 2000;320:1373–8.

43. Wagner EH, Sandhu N, Newton KM, et al. Effect of improved glycemic control on healthcare costs and utilization. JAMA 2001;285:182–9.

44. Price CP, St John A. Managing diabetes in the community setting. Point-of-care testing. 2006 (accepted for publication).

45. Patsalos PN, Sander JWAS, Oxley J, et al. Immediate anticonvulsive drug monitoring in management of epilepsy. Lancet 1987;2:39.

46. Elliot K, Watson ID, Tsintis P, et al. The impact of near patient testing on the organization and costs of an anticonvulsant clinic. Therap Drug Monit 1990;120:434–7.

47. Rohlfing CL, Wiedmeyer HM, Little RR, et al. Defining the relationship between plasma glucose and HbA1c: analysis of glucose profiles and HbA1c in the Diabetes Control and Complications Trial. Diabetes Care 2002;25:275–8.

48. Agency for Health Research and Quality. Evidence report/technology assessment: number 84. Use of glycated hemoglobin and microalbuminuria in the monitoring of diabetes mellitus. Available at http://www.ahrq.gov/clinic/evrptfiles.htm#glyca (Accessed February 12, 2006).

49. British Medical Association. Quality and outcomes framework guidance. Investing in general practice. The new general medical services contract. Annex A: quality indicators—summary of points. Available at http://www.bma.org.uk/ap.nsf/Content/NewGMSContract/$file/gpcontractannexa.pdf (Accessed February 12, 2006).

50. Goldstein DE, Little RR, Lorenz RA, et al. Tests of glycemia in diabetes. Diabetes Care 2004;27 Suppl 1:S91–3.

51. Department of Health. National service framework for diabetes. Available at http://www. publications.doh.gov.uk/nsf/diabetes/index.htm (Accessed February 12, 2006).

52. National Institute for Clinical Excellence. Type 2 diabetes - management of blood glucose. Available at http://www.nice.org.uk/page.aspx?o=36881 (Accessed February 12, 2006).

53. Canadian Diabetes Association. Clinical Practice Guidelines Committee. Clinical practice guidelines 2003 for the prevention and management of diabetes in Canada 2003. Available at http://www.diabetes.ca/cpg2003/chapters.aspx (Accessed February 12, 2006).

54. Miedema K. Laboratory tests in diagnosis and management of diabetes mellitus. Practical considerations. Clin Chem Lab Med 2003;41:1259–65.

55. Klein R, Klein BE, Moss SE, et al. The 10-year incidence of renal insufficiency in people with type 1 diabetes. Diabetes Care 1999;22:743–51.

56. Stratton IM, Adler AI, Neil HA, et al. Association of glycaemia with macrovascular and microvascular complications of type 2 diabetes (UKPDS 35): prospective observational study. BMJ 2000;321:405–12.

57. St John A, Davis TM, Goodall I, et al. Nurse-based evaluation of point-of-care assays for glycated haemoglobin. Clin Chim Acta 2006;365:257–63.

58. ECRI Reviews Glycohemoglobin Analyzers. Available at http://www.ecri.org/Newsroom/Document_Detail.aspx?docid=20031219_97 (Accessed February 12, 2006).

59. Cagliero E, Levina E, Nathan D. Immediate feedback of HbA1c levels improves glycemic control in type 1 and insulin-treated type 2 diabetic patients. Diabetes Care 1999;22:1785–9.

60. Thaler LM, Ziemer DC, Gallina DL, et al. Diabetes in urban African-Americans. XVII. Availability of rapid HbA1c measurements enhances clinical decision-making. Diabetes Care 1999;22:1415–21.

61. Miller CD, Barnes CS, Phillips LS, et al. Rapid A1c availability improves clinical decision-making in an urban primary care clinic. Diabetes Care 2003;26:1158–63.

62. Kennedy L, Herman WH, Strange P, et al. Impact of active versus usual algorithmic titration of basal insulin and point-of-care versus laboratory measurement of HbA1c on glycemic control in patients with type 2 diabetes: the Glycemic Optimization with Algorithms and Labs at Point of Care (GOAL A1C) trial. Diabetes Care 2006;29:1–8.

63. Grieve R, Beech R, Vincent J, Mazurkiewicz J. Near patient testing in diabetes clinics: appraising the costs and outcomes. Health Technol Ass 1999; 3:1–74.

64. Ferenczi A, Reddy K, Lorber DL. Effect of immediate hemoglobin A1c results on treatment decisions in office practice. Endocr Pract 2001;7:85–8.

65. Holman RR, Jelfs R, Causier PM, et al. Glycosylated haemoglobin measurement on blood samples taken by patients: an additional aid to assessing diabetic control. Diabet Med 1987;4:71–3.

66. Heisler M, Piette JD, Spencer M, et al. The relationship between knowledge of recent HbA1c values and diabetes care understanding and self-management. Diabetes Care 2005;28:816–22.

67. Levetan CS, Dawn KR, Robbins DC, et al. Impact of computer-generated personalized goals on HbA1c. Diabetes Care 2002;25:2–8.

68. Oral Anticoagulation Monitoring Study Group. Point-of-care prothrombin time measurement for professional and patient self-testing use. A multicenter clinical experience. Oral Anticoagulation Monitoring Study Group. Am J Clin Pathol 2001;115:288–96.

69. Shiach CR, Campbell B, Poller L, et al. Reliability of point-of-care prothrombin time testing in a community clinic: a randomized crossover comparison with hospital laboratory testing. Br J Haematol 2002;119:370–5.

70. Sawicki PT. A structured teaching and self-management program for patients receiving oral anticoagulation: a randomized controlled trial. Working Group for the Study of Patient Self-Management of Oral Anticoagulation. JAMA 1999;281:145–50.
71. Fitzmaurice DA, Hobbs FD, Murray ET, et al. Evaluation of computerized decision support for oral anticoagulation management based in primary care. Br J Gen Pract 1996;46:533–5.
72. Fitzmaurice DA, Hobbs FD, Murray ET, et al. Oral anticoagulation management in primary care with the use of computerized decision support and near-patient testing: a randomized, controlled trial. Arch Intern Med 2000;160:2343–8.
73. Poller L, Shiach CR, MacCallum PK, et al. Multicentre randomised study of computerised anticoagulant dosage. European Concerted Action on Anticoagulation. Lancet 1998;352:1505–9.
74. Manotti C, Moia M, Palareti G, et al. Effect of computer-aided management on the quality of treatment in anticoagulated patients: a prospective, randomized, multicenter trial of APROAT (Automated Program for Oral Anticoagulant Treatment). Haematologica 2001;86:1060–70.
75. Kovacs MJ, Anderson DA, Wells PS. Prospective assessment of a nomogram for the initiation of oral anticoagulation therapy for outpatient treatment of venous thromboembolism. Pathophysiol Haemost Thromb 2002;32:131–3.
76. Siebenhofer A, Berghold A, Sawicki PT. Systematic review of studies of self-management of oral anticoagulation. Thromb Haemost 2004;91:225–32.
77. Heneghan C, Alonso-Coello P, Garcia-Alamino JM, et al. Self-monitoring of oral anticoagulation: a systematic review and meta-analysis. Lancet 2006; 367:404–11.
78. Sidhu P, O'Kane HO. Self-managed anticoagulation: results from a two-year prospective randomized trial with heart valve patients. Ann Thorac Surg 2001;72:1523–7.
79. Koertke H, Minami K, Boethig D, et al. INR self-management permits lower anticoagulation levels after mechanical heart valve replacement. Circulation 2003;108 Suppl 1:1175–8.
80. Christensen TD, Attermann J, Pilegaard HK, et al. Self-management of oral anticoagulant therapy for mechanical heart valve patients. Scand Cardiovasc J 2001;35:107–13.
81. Kortke H, Korfer R. International normalized ratio self-management after mechanical heart valve replacement: is an early start advantageous? Ann Thorac Surg 2001;72:44–8.
82. Valenstein P. Can we satisfy clinicians' demands for faster service? Should we try? Am J Clin Pathol 1989;91:705–6.
83. Parvin CA, Lo SF, Deuser SM, et al. Impact of point-of-care testing on patients' length of stay in a large emergency department. Clin Chem 1996;42:711–7.
84. Kendall J, Reeves B, Clancy M. Point-of-care testing: randomised, controlled trial of clinical outcome. BMJ 1998;316:1052–7.
85. Murray RP, Leroux M, Sabga E, et al. Effect of point-of-care testing on length of stay in an adult emergency department. J Emer Med 1999;17:811–4.
86. Lee-Lewandrowski E, Corboy D, Lewandrowski K, et al. Implementation of a point-of-care satellite laboratory in the emergency department of an academic medical center. Impact on test turnaround time and patient emergency department length of stay. Arch Pathol Lab Med 2003;127:456–60.

87. Ng SM, Krishnaswamy P, Morissey R, et al. Ninety-minute accelerated critical pathway for chest pain evaluation. Am J Cardiol 2001;88:611–7.

88. McCord J, Nowak RM, McCullough PA, et al. Ninety-minute exclusion of acute myocardial infarction by use of quantitative point-of-care testing of myoglobin and troponin I. Circulation 2001;104:1483–8.

89. Zarich S, Bradley K, Seymour J, et al. Impact of troponin T determinations of hospital resource utilization and costs in the evaluation of patients with suspected myocardial ischemia. Am J Cardiol 2001;88:732–6.

90. Ohman EM, Armstrong PW, White HD, et al. Risk stratification with a point-of-care cardiac troponin T test in acute myocardial infarction. Am J Cardiol 1999;84:1281–6.

91. Kline JA, Israel EG, Michelson, EA, et al. Diagnostic accuracy of a bedside D-dimer assay and alveolar dead-space measurement for rapid exclusion of pulmonary embolism. JAMA 2001;285:761–8.

92. Bates SM, Grand'Maison A, Johnston M, et al. A latex D-dimer reliably excludes venous thromboembolism. Arch Intern Med 2001;161:447–53.

93. Brown MD, Lau J, Nelson RD, Kline JA. Turbidimetric D-dimer test in the diagnosis of pulmonary embolism: a meta-analysis. Clin Chem 2003;49:1846–53.

94. Kline JA, Webb WB, Jones AE, Hernandez-Nino J. Impact of a rapid rule-out protocol for pulmonary embolism on the rate of screening, missed cases, and pulmonary vascular imaging in an urban US emergency department. Ann Emerg Med 2004;44:490–502.

95. Roy PM, Colombet I, Durieux P, Chatellier G, Sors H, Meyer G. Systematic review and meta-analysis of strategies for the diagnosis of suspected pulmonary embolism. BMJ 2005;331:259–68.

96. Seaman MJ, Culank LS, Price CP. Perioperative laboratory support. In: Calne RY, ed. Liver transplantation, 2nd ed. London: Grune and Stratton, 1988:199–208.

97. Despotis GJ, Joist JH, Hogue CW, et al. The impact of heparin concentration and activated clotting time monitoring on blood conservation: a prospective, randomized evaluation in patients undergoing cardiac operations. J Thorac Cardiovasc Surg 1995;110:46–54.

98. Despotis GJ, Joist JH, Goodnough LT. Monitoring of hemostasis in cardiac surgical patients: impact of point-of-care testing on blood loss and transfusion outcomes. Clin Chem 1997;43:1684–96.

99. Irvin GL, Molinari AS, Figueroa C, et al. Improved success rate in reoperative parathyroidectomy with intraoperative PTH assay. Ann Surg 1999;229:874–9.

100. Irvin GL, Solorzano CC, Carneiro DM. Quick intra-operative parathyroid hormone assay: surgical adjunct to allow limited parathyroidectomy, improve success rate, and predict outcome. World J Surg 2004;28:1287–92.

101. Chen H, Sokoll LJ, Udelsman R. Outpatient minimally invasive parathyroidectomy: a combination of sestamibi-SPECT localization, cervical block anesthesia, and intra-operative parathyroid hormone assay. Surgery 1999;126:1016–22.

102. Walsh TS, Dodds S, McArdle F. Evaluation of simple criteria to predict successful weaning from mechanical ventilation in intensive care patients. Br J Anaesth 2004;92:793–9.

103. Halpern MT, Palmer CS, Simpson KN, et al. The economic and clinical efficiency of point-of-care testing for critically ill patients: a decision-analysis model. Am J Med Qual 1998;13:3–12.

104. Kilgore ML, Steindel SJ, Smith JA. Cost analysis for decision support: the case of comparing centralized versus distributed methods for blood gas testing. J Healthcare Management 1999;44:207–15.
105. Rossi AF, Khan DM, Hannan R, et al. Goal-directed medical therapy and point-of-care testing improve outcomes after congenital heart surgery. Intensive Care Med 2005;31:98–104.
106. Vanhorebeek I, Langouche L, Van den Berghe G. Glycemic and nonglycemic effects of insulin: how do they contribute to a better outcome of critical illness? Curr Opin Crit Care 2005;1:304–11.
107. Van den Berghe GH. Role of intravenous insulin therapy in critically ill patients. Endocr Pract 2004;10 Suppl 2:17–20.
108. Van den Berghe G, Wouters PJ, Bouillon R, et al. Outcome benefit of intensive insulin therapy in the critically ill: insulin dose versus glycemic control. Crit Care Med 2003;31:359–66.
109. Grey NJ, Perdrizet GA. Reduction of nosocomial infections in the surgical intensive-care unit by strict glycemic control. Endocr Pract 2004;10 Suppl 2:46–52.
110. Krinsley JS. Effect of an intensive glucose management protocol on the mortality of critically ill adult patients. Mayo Clin Proc 2004;79:992–1000.
111. Collier B, Diaz J, Forbes R, et al. The impact of a normoglycemic management protocol on clinical outcomes in the trauma intensive care unit. JPEN-Parenter-Enter 2005;29:353–8.
112. Plank J, Blaha J, Cordingley J, et al. Multicentric, randomized, controlled trial to evaluate blood glucose control by the model predictive control algorithm versus routine glucose management protocols in intensive care unit patients. Diabetes Care 2006;29:271–6.
113. Kanji S, Buffie J, Hutton B, et al. Reliability of point-of-care testing for glucose measurement in critically ill adults. Crit Care Med 2005;33:2778–85.
114. Allison KP, Gosling P, Jones S, et al. Randomized trial of hydroxyethyl starch versus gelatin for trauma resuscitation. Trauma 1999;47:1114–21.
115. Gosling P, Czyz J, Nightingale P, Manji M. Factors that determine microalbuminuria on admission to intensive care in 431 medical and surgical patients. Crit Care Med 2006 (accepted for publication).

Index